South American Cichlids IV
Discus & Scalare

Manfred Göbel
Hans J. Mayland

Haftung:
Alle Angaben in diesem Buch sind nach bestem Wissen und Gewissen niedergeschrieben.
Für eventuelle Fehler schließen die Autoren und der Verlag jegliche Haftung aus. Mit dem Erwerb dieses Buches erkennt der Eigentümer diesen Haftungsausschluß ausdrücklich an.

Liability:
All the information in this book has been recorded with due care and attention.
The authors and the publishers will not accept liability for any inaccuracies.
By purchasing this book the owner explicitly accepts this disclaimer of liablity.

Die Deutsche Bibliothek - CIP-Einheitsaufnahme

Aqualog : reference fish of the world. - Mörfelden-Walldorf : A.C.S.
South American Cichlids. IV (Diskus , Scalare).-1998

South American Cichlids / Manfred Göbel; Hans J. Mayland. - Mörfelden-Walldorf : A.C.S. (Aqualog)

Bis Bd. 2 verf. von Ulrich Glaser sen. und Wolfgang Glaser
Bd. 3 verf. von Ulrich Glaser sen., Frank Schäfer und Wolfgang Glaser

NE: Göbel, Manfred; Mayland, Hans J.
4 (1998)

ISBN 3-931702-75-8

© Copyright: Verlag A.C.S. GmbH

Rothwiesenring 5, 64546 Mörfelden-Walldorf / Germany

Alle Rechte vorbehalten, Reproduktion, Speicherung in Datenverarbeitungsanlagen, Wiedergabe auf elektronischen, fotomechanischen oder ähnlichen Wegen, Funk und Vortrag - auch auszugsweise - nur mit Genehmigung des Verlages.

All rights reserved. No part of this publication may be reproduced or transmitted, in any form or by any means, without permission.

Text und fachliche Bearbeitung:
Frank Schäfer, Michael Kempkes
Übersetzungen (außer/*exept*: pp. 225-231)
Monika Schäfer
Index und Organisation
Wolfgang Glaser
Redaktion
Frank Schäfer
Titelgestaltung:
Gaby Geiß, Büro für Grafik, Frankfurt a.M.
Druck, Satz, Verarbeitung:
Lithos Verlag A.C.S.
Bildbearbeitung Frank Teigler
Giese Druck, Offenbach
Gedruckt auf EURO ART glänzend,
100% chlorfrei von PWA Umweltfreundlich

Printed in Germany

Inhalt
Contents

Diskus und Skalare; Zwei Buntbarschgattungen, deren Mitglieder eine erstaunliche aquaristische Karriere machten *	Seite 4
Discus and Angels: Two Cichlid Genera with a Career	*page 4*
Zur Artenfrage bei Diskus und Skalaren; Anmerkungen aus biologischer Sicht **	Seite 22
About the species classification; from a biological point of view	*page 22*
Südostasien - Europa; Unterschiede in der Diskuszucht ***	Seite 29
Southeast Asia and Europe ; Differences in Commercial Discus Breeding	*page 29*
Symbolerklärung deutsch	Seite 32
Explanation of the symbols, german	*page 32*
Bildteil Diskus ***	Seite 33
Plates Discus	*page 33*
Bildteil Wildformen Skalare **	Seite 179
Plates Angels, wild forms	*page 179*
Bildteil Zuchtformen Skalare ****	Seite 193
Plates Angels, breeding forms	*page 193*
Diskus - Bewertung ***	Seite 225
Assessment of Discus	*page 225*
Index Code-Nummern ****	Seite 232
Index Code-numbers	*page 232*
Cichliden-Verbände weltweit	Seite 238
Cichlid Associations worldwide	*page 238*
Symbolerklärung englisch	Seite 240
Explanation of the symbols, english	*page 240*

* by H.-J. Mayland ** by F. Schäfer *** by M. Göbel **** by U. Glaser sen.

Erklärungen der Abkürzungen in den wissenschaftlichen Namen
Key to the abbreviations of the scientific names

Beispiel/	*Symphysodon*	*aequifasciatus*	*aequifasciatus*	PELLEGRIN,	1904
example:	Gattung	Art	Unterart	Beschreiber	Publikationsjahr
	Genus	*Species*	*Subspecies*	*Describer*	*Year of publication*

sp./spec.: = species (lat.): Art / *species*
Hinter einem Gattungsnamen meint dies: Ein Artname steht (noch) nicht zur Verfügung, die Art ist bislang nicht eindeutig bestimmt bzw. noch nicht formell beschrieben./
Following the genus name, this means: A species name is not yet available, the species has not yet been determined or formally described.

ssp.: = subspecies (lat.): Unterart / *subspecies*
Einige Arten haben ein sehr großes Verbreitungsgebiet; innerhalb dieses Gebietes gibt es Populationen, die sich äußerlich zwar deutlich von anderen Populationen unterscheiden, genetisch jedoch zur gleichen Art gehören. Solche Populationen erhalten als Unterart einen wissenschaftlichen Namen. Ist die Unterart bislang unbenannt, so steht hier nur ssp./
Some species inhabit an area of a very wide range; within this area, there may be populations which differ significantly in appearance from other populations, but clearly belong to the same species. Such populations may get a third scientific name, indicating a subspecies. If a subspecies name has not yet been formally given, the abbreviation ssp. is added.

cf.: = confer (lat.): vergleiche / *compare*
Einem Artnamen vorangestellt meint dies: Das vorliegende Exemplar oder die entsprechende Population weicht in gewissen Details von der typischen Form ab, jedoch nicht so gravierend, daß es oder sie einer anderen Art zugeordnet werden könnte./
Placed in front of a species name this means: The specimen shown or the respective population to which it belongs differ in some minor details from the typical form, but these differences do not justify to place it into a species of its own.

sp. aff.: = species affinis (lat.): ... ähnliche Art / *similar to ...*
Einem Artnamen vorangestellt meint dies: Die vorliegende Art ist bisher noch nicht bestimmt, sie ähnelt jedoch der genannten und bereits beschriebenen Art./
Placed in front of a species name, this means: The species at hand is not yet determined but it is very similar to the one named in the following.

Hybride/*hybrid*: Kreuzungsprodukt, Mischling zweier Arten / *hybrid or crossbreed of two species*

leg. : = legat (lat.): Überbringer, Bote / *a person who brings or gives something*
Im AQUALOG - Kontext bedeutet dies: die exakte Fundortangabe zu dem abgebildeten Fisch stammt von der in Kapitälchen geschriebenen Person / *This means in the AQUALOG context: the information about the exact place of origin of the fish depicted is given by the person written in capitals*

Discus and Angels - two Cichlid Genera with a Career

By Hans J. Mayland

Introduction

When in 1840, over one and a half centuries ago, Jacob HECKEL classified nine new genera, the genera Symphysodon and Pterophyllum were among them (just like Uaru, Acara, Heros, Geophagus, Chaetobranchus, Crenicichla and Batrachops).

With HECKEL's classification, Symphysodon discus was the first ever described discus. On the other hand, the Angel had been first described several years earlier: In 1823, LICHTENSTEIN described Zeus scalaris. In his 1840 work, HECKEL introduced Pterophyllum as a valid genus and assigned P. scalaris as genus type. In those days, the species name was chosen regardless of the sex of the genus name; this is why the name of this fish should be scalare, according to the neutral sex of Pterophyllum (which comes from the Greek pteron = fin, wing, sail and phyllum = leaf, so that the whole name means 'like a winged leaf') and not scalaris that was given by HECKEL and has a masculine ending.

The history of Jacob HECKEL and his work is probably well-known to most enthusiasts; MAYLAND gave a thorough account in his "Adventures with Discus'" (T.F.H. Publications Inc., Neptune City, USA, 1994). Still, I would like to summarise shortly the most important stages of the life and work of this remarkable man. At the beginning of the 19th century, Johann Jacob HECKEL indulged himself in fieldwork, but his research was limited to European species. After a 'theoretical detour' ("The Fishes of Kashmir", 1819), he started 'genuine' research in the field of tropical fishes with his 1840 book. Johann NATTERER who came back to Austria in 1836 after eighteen years in Brazil put his fish collection at HECKEL's disposal - and, at last, Jacob HECKEL had material he could truly work with.

But how were the two researchers connected with Austria's imperial monarchy and its natural history cabinet?

Heckels Leben hatte viele Stationen, die auf die politische Lage der damaligen Zeit zurückzuführen sind: Seine Eltern waren lange Zeit auf der Flucht. Geboren wurde J. J. Heckel am 23. Januar 1790 in Mannheim, wo sein Vater als Kapellmeister sein Brot verdiente. Doch bereits drei Jahre später begann der lange Fluchtweg, der die Familie über die Schweiz, Regensburg und Wien nach Preßburg führte. Der junge Heckel wurde zunächst im Haus der Eltern unterrichtet, kam jedoch 1799 — wieder zurück in Mannheim — in das dortige kurfürstliche Institut. Aus dem Studium von Büchern lernte er die Natur kennen, und seine Liebe zu den Naturwissenschaften erwachte. Im Jahre 1801 kehrte Heckel nach Wien zurück. Hier waren seine Eltern in der Zwischenzeit seßhaft geworden, und ihr Sohn begann mit den Vorstudien zu einer landwirtschaftlichen Ausbildung. Im Jahre 1806, wieder einmal auf der Flucht, trat der inzwischen Sechzehnjährige in Preßburg in die vom Grafen Georg Festetics in Keszthely errichtete landwirtschaftliche Georgicon-Schule ein, um sich hier als Landwirt ausbilden zu lassen. Nach dem Abschluß an dieser Schule im Jahre 1909 begann der junge Landwirt auf dem von den Eltern inzwischen erworbenen Besitz in Gumpoldskirchen die Landwirtschaft praktisch auszuüben. Nebenher befaßte er sich weiter mit Botanik, sammelte Vogelbälge und begann autodidaktisch — also ohne weitere Anleitung — diese Bälge zu präparieren. Durch diese Arbeit bekam er schließlich Kontakt mit dem k. k. Hof-Naturalienkabinett. Dabei lernte er den damaligen Kustos Joseph Natterer kenne, den Vater von Johann.

Nach einigen weiteren Jahren und den sich ständig erweiternden Kontakten wurde die Liebe zu den Naturwissenschaften in Heckel so groß, daß er die Landwirtschaft aufgab, nach Wien übersiedelte und sich hier im Präparieren ausbilden ließ. Um seine bescheidenen finanziellen Ansprüche befriedigen zu können, begann er einen Handel mit den zu jener Zeit begehrtenVogelbälgen. Das Geschäft lief mehr als nur befriedigend, so daß er sich bald eine erste Auslandreise nach Italien leisten konnte, von der er neben einer allgemein gehaltenen Sammlung auch eine Anzahl von Meeresfischen mitbrachte. Er übergab alles dem k. k. Hof-Naturalienkabinett, für das er darauf zu einem damals üblichen Tageslohn (!) die Arbeit eines Präparators übernehmen durfte. Zu jener Zeit war die Fischsammlung dort noch mehr als dürftig, wohingegen an anderen namhaften Plätzen bisher wesentlich intensiver an dieser Materie gearbeitet worden war. Heckel legte daraufhin sein Augenmerk auf die Fische und erarbeitete sich weitere Kenntnisse. Als glücklichen Umstand kann man die Tatsache ansehen,

Heckel's life ran through many stations that were closely connected to the political situation his parents found themselves in: For a very long time, they lived as refugees. Jacob Heckel was born on the 23 January 1790, in Mannheim/Germany where his father earned his money being a director of music. But already three years later the refugee life began when the parents first fled to Switzerland, then to Regensburg, Vienna, and finally to Preßburg. Young Jacob Heckel was educated in his parents' house, but in 1799 he returned to Mannheim and joined the electoral institute. There, he read many books about nature and "fell in love" with the natural sciences. In 1801 he returned to Vienna where his parents had finally settled down and started his agricultural training. Five years later, again on the run from political disturbances, the sixteen-year-old Jacob Heckel joined the agricultural training school of Count Georg Festetics in Keszthely, the so-called 'Georgicon-School' in Preßburg. In 1909 he graduated from this school and started to work as a farmer on a piece of land near Gumboldtskirchen that had been bought by his parents. Apart from his duties as a farmer he kept on studying botany and tought himself how to preserve the dead birds he collected. The skill to preserve dead birds finally lead to the contact with the imperial natural history cabinet and its present director, Joseph Natterer, Johann's father.

Some years later, Heckel's love for nature and the natural sciences had grown so much that he decided to give up farming. He moved to Vienna and trained as a preserver. To be able to finance his modest lifestyle he started to sell his bird skins which were then much sought-after objects. His business was more than successful and soon he could afford to travel to Italy where he not only collected animals of general interest but also some marine fishes. Heckel gave the whole collection to the imperial natural history cabinet that hired him to preserve the collection, paying him the then usual daily (!) wages. In those days the fish collection was more than poor whereas in other renowned towns the research in this special field was pursued much more intensely. Consequently, Heckel concentrated on fishes and acquired much knowledge about them. He was also really lucky, because the famous Swiss biologist Louis Agassiz visited Vienna in 1830 and stayed for a while so that Heckel found himself in the most fortunate position to work with a man who could teach him so very much. As

daß der hochangesehene schweizer Zoologe Louis AGASSIZ 1830 Wien für eine längere Zeit besuchte und HECKEL mit ihm zusammenarbeiten und weiter lernen konnte. Es folgten entsprechende Arbeiten, und er trat mit Werken an die Öffentlichkeit, die seinen Ruf als Systematiker begründeten. Als 1836 Johann NATTERER aus Brasilien zurückkehrte, befanden sich in seinem umfangreichen Gepäck auch viele Fischpräparate, die für HECKEL die Basis seines 1840 publizierten Werkes »Johann Natterers neue Flussfische Brasilien's; nach Beobachtungen des Entdeckers beschrieben« darstellten.

HECKEL hat, wie man erfahren konnte, an keiner Universität studiert, geschweige denn promoviert. Obgleich er als guter Zeichner bekannt war, hat er seiner Diskus-Beschreibung (es gab davon damals nur 1 Präparat!) keine Zeichnung beigefügt. Sie erfolgte erst Jahre später, als Dr. Rudolf KNER sich erneut dieser Art in einer Arbeit annahm.

Johann Jacob HECKEL stieg bei der Reorganisation des k. k. Hof-Naturalienkabinetts einen Rang nach oben, und im Jahre 1851 wurde ihm die Oberaufsicht über die ornithologische und ichthyologische Sammlung unterstellt. In den Jahren danach begann er zu kränkeln und starb schließlich, am 1 März 1857, im Alter von 67 Jahren nach einem arbeitsreichen und (zumindest aus unserer heutigen Sicht) erfolgreichem Leben. Mit Sicherheit hätte sich HECKEL nicht träumen lassen, nach mehr als 150 Jahren namentlich in aller Munde zu sein.

Johann NATTERER wurde 1787 in Laxenburg/Österreich als Sohn des Joseph N. geboren. Der damalige österreichische Kaiser Franz, der Gefallen an den Vogelpräparaten seines ›berittenen Falkenjägers‹ Joseph bekundete, holte diesen mit seiner Familie nach Wien und bestellte ihn zum Aufseher seiner neuen Sammlung. Seinem Sohn Johann bekam die Hauptstadt gut, und er besuchte an der Realakade-mie Kurse für Zeichnen und Fremdsprachen, wobei er sich beim Zeichnen zu einer wahren Meisterschaft steigerte. Dazu kannte er sich auch gut mit dem Präparieren aus, so daß er schließlich an Vorlesungen der Universität über Anatomie, Botanik und Chemie an der Wiener Hochschule teilnahm. Die Bezahlung für jegliche Arbeit war zu dieser Zeit jedoch sehr schlecht. Wegen seiner vielen Kenntnisse wurde gerade er jedoch eingeladen, im Rahmen der großen Brasilien-Expedition des Kaisers Franz, die dieser 1817 anläßlich der Vermählung seiner Tochter Leopoldine mit dem Kronprinzen von Brasilien, Don Pedro, ausrüstete,

a result of this fruitful co-operation, HECKEL published several books which established him as a competent systematist. When in 1836 Johann NATTERER returned from Brazil he brought with him a large collection of preserved specimens of many fish species. This collection was the basis of HECKEL 1840 work "Johann Natterers Flussfische Brasiliens; nach Beobachtungen des Entdeckers beschrieben" (Johann Natterer's River Fishes of Brazil; described after the Discoverer's Observations).

As one can gather from what was said about HECKEL'S life, he never studied biology or even did a PhD degree. Although he was known to be a good draughtsman, his description of the discus (there was only one single preserved specimen!) did not enclose a drawing of the fish. This was finally done years later, when Dr. Rudolf KNER re-examined the species for one of his works.

When the imperial natural history cabinet was reorganised, Jacob HECKEL was promoted and in 1851 he was in charge of the ornithological and the ichthyological collection. Unfortunately, a little later he started to suffer from poor health and on 1 March 1857 he died at the age of 67, having lead a busy and (at least from today's point of view) successful life. However: One can be quite sure that HECKEL would have never dreamt of being - 150 years later - one of the hobby's 'celebrities'.

Johann NATTERER *was born in 1787 in Laxenburg/Austria. His father Joseph NATTERER was one of Emperor Franz' huntsmen. The Austrian Emperor was a huge fan of Joseph NATTERER'S preserved birds and he appointed him as director of the new Viennese collection. Joseph NATTERER moved with his family to Vienna and his son Johann flourished under the influence of this new, exciting life in Austria's capital. He visited drawing and foreign language classes and became a real 'master' draughtsman. Moreover, because of his father's work he knew quite a lot about the preservation of animal corpses, so that he finally attended classes at Vienna University: anatomy, botany, and chemistry. In the nineteenth century, all work was paid poorly, but Johann NATTERER was fortunate to know a lot about biology and animal preservation, so that when in 1817 Emperor Franz set up a research expedition to Brazil in honour of his daughter's marriage with Brazil's Prince Dom Pedro, he appointed young Johann NATTERER to be*

als Fachexperte das Gesamtgebiet der Zoologie zu übernehmen — eine glänzende Bewährungsprobe!

Als NATTERER schließlich nach 18 Jahren aus Brasilien zurückkehrte, konnte man feststellen, daß der mitgebrachte Rest der Schätze, von denen ihm ein großer Teil gestohlen worden war, keinesfalls mühelos erworben worden war. Er hatte dafür viele Opfer bringen müssen und war zum Beispiel immer wieder von tückischen Fiebern in der ›Grünen Hölle‹ niedergeworfen worden. Er verlor bereits nach 4 Jahren Aufenthalt in Brasilien seinen treuen Begleiter SOCHOR, als sie den bis dahin noch von keinem Forscher betretene Mato Grosso bereisten, dessen Hauptstadt gleichen Namens (heute ›Vila Bela da Santissima Trinidade‹) nahe der Grenze zu Bolivien sie im Oktober 1826 erreichten. Später (1831) vermählte sich NATTERER in Barcelos am Rio Negro (woher heute die meisten Roten Neonfische kommen) mit einer Brasilianerin. Als 1835 ein heftiger Bürgerkrieg in Brasilien tobte, verließ NATTERER mit seiner Famile das Land. Nach Österreich zurückgekommen, mußte er leider feststellen, daß sich an der miesen finanziellen Lage, verbunden mit einer Günstlingswirtschaft, nicht viel geändert hatte. Seine mitgebrachte Fischsammlung allein umfaßte 1.700 Exemplare von insgesamt 50.000 sorgfältig konservierten Präparaten (darunter 13.300 Vögel).

Jonann NATTERER bereiste in den folgenden Jahren viele Museen im Westen und Süden Europas, bei denen er feststellte, daß das Ausland seine Verdienste mehr würdigte als das eigene Land. So verlieh ihm die damals schon geachtete Universität von Heidelberg ein Ehrendoktorat der Philosophie. Die Wiener Universität hat leider geschwiegen! Mitten in seinem rastlosen Arbeiten wurde NATTERER von einem Blutsturz überrascht. Er verstarb am 17. Juni 1843 im 56. Lebensjahr.

Diskusfische

Jacques PELLEGRIN, der französische Ichthyologe, war nicht nur ein bedeutender Wissenschaftler seines Landes: Er hatte auch aquaristische Interessen!

Der besondere Verdienst des Autors in Bezug auf die Diskusfische ist die aus dem Jahre 1903 datierte Beschreibung der zweiten Diskus-Art, nämlich *Symphysodon aequifasciatus*, die PELLEGRIN zwar zunächst als Variante von *S. discus* ansah, von späteren Ichthyologen jedoch zunächst in den Rang einer Unterart und darauf

the head of the zoological part of this expedition. What a fantastic chance! Eighteen years later, when Johann NATTERER returned from South America, he could only present fragments of the treasures he had collected in Brazil's 'green hell'. Most of the things he wanted to bring to his home country had been stolen: but this fact was not as dramatic as the high price he had to pay in the name of the Emperor. Apart from frequent, bad fever attacks he had to endure in the rainforest, he had to cope with the death of his dedicated comrade Sochor who died in 1821 when the expedition reached the until then by no foreign scientist examined Mato Grosso.

Five years later they arrived at the capital of the same name (which is today called 'Vila Bela da Santissima Trinidade') that is located near the Bolivian border. At the Rio Negro (which is famous today as being the main source of Cardinal Tetras), NATTERER married in 1831 a Brazilian woman. In 1835 a terrible Civil War broke out in Brazil, and NATTERER was forced to leave the country, taking his family with him to Austria. Having come back to his home country after 18 years abroad, Johann NATTERER had to realise that in terms of poor pay and favouritism nothing had changed. Although his fish collection alone included 1700 preserved specimens of a total collection of 50 000 (including 13 300 birds) preserved animals, he never even was honoured by the University of Vienna.

In the following years, NATTERER travelled to many countries of western and southern Europe and visited museums, only to find out that in foreign countries, his work was much more honoured than in Austria. He even received an Honour Doctorate of Philosophy of the (then already famous) University of Heidelberg. In the middle of restless work, on 17 June 1843, Johann NATTERER died of an unexpected haemorrhage. He was 56 years old.

DISCUS

Jacques PELLEGRIN, the French ichthyologist, was not only one of the most important scientists of his country, he also was an enthusiastic hobbyist!

In regard of the discus, he holds a special place, because in 1903, PELLEGRIN described the second species of the discus, *Symphysodon aequifasciatus*. When describing the

in den einer selbständigen Art erhoben wurde.

Im Jahre 1960 überarbeitete der US-amerikanische Ichthyologe Leonard P. SCHULTZ die Gattung *Symphysodon* und stellte zu *S. aequifasciatus* zwei neue Unterarten auf, *S. a. axelrodi* und *S. a. haraldi*, wodurch automatisch von der *S. aequifasciatus* die Nominat-Unterart *S. a. aequifasciatus* zu bilden ist, so daß im Grunde nun drei Unterarten existent sind (Art. 47a der Internationalen Nomenklaturregeln).

Eine weitere Unterart wurde 1981 von US-amerikanischen Ichthyologen Warren E. BURGESS von *S. discus* mit *S. d. willischwartzi* aufgestellt, worauf wieder eine Nominat-Unterart zu bilden war: *S. discus discus*.

Der schwedische Ichthyologe Sven O. KULLANDER untersuchte Exemplare von *S. aequifascuatus* und veröffentlichte die Ergebnisse 1986 in ›Cichlid fishes of the Amazon River drainage of Peru‹ unter Bezugnahme auf die Unterarten von SCHULTZ: »I do not find any significant differences. —I cannot repeat SCHULTZ's counts. —The squamation is doubtful.«.... (»Ich kann keine kennzeichnenden Farbunterschiede feststellen. —Ich kann SCHULTZ' Schuppenzählung nicht nachvollziehen. —Die Zahlen sind zweifelhaft«)....und setzte beide Unterarten in die Synonymie! Für *S. d. willischwartzi* dürfte Ähnliches gelten, doch zu diesen Fischen nahm KULLANDER in diesem Buch nur am Rande Stellung, wohl weil *S. discus* im Rio Negro beheimatet ist.

»Gibt es eine oder zwei Arten Diskusfische?« wird oft gefragt. Diese Frage konnte bisher von keinem Wissenschaftler klar beantwortet werden. Zwar haben sich verschiedene Züchter zum Ziel gesetzt, Tiere von *S. discus* und *S. aequifasciatus* zu kreuzen. Das ist auch gelungen, doch haben diese Fische weder die Schönheit der einen noch der anderen Art. Daß eine Kreuzung überhaupt möglich ist, ist kein Beleg dafür, daß die derzeit zwei wissenschaftlich beschriebenen Taxa zu einem zusammengefaßt werden müßten (siehe hierzu auch die Anmerkungen „Zur Artenfrage", S. 22). Eine Bastardisierung kann zuweilen aus aquaristischer Sicht etwas Positives zustande bringen. Bei den Diskusfischen mit dem dunklen, ungemusterten Mittelbalken ist das jedoch nach der Meinung der meisten Aquarianer nicht gelungen.

Die Verbreitung der Diskusfische reicht über das gesamte zentrale Amazonasbecken Brasiliens sowie in Teile des Rio Putumayo (Peru) hinein. Die aus dem Río

fish for the first time, PELLEGRIN thought it to be a variety of the first species, S. discus. But ichthyologists who re-examined the fish later on classified it first as subspecies of S. discus, and after some time it got finally the status of a valid, independent species. In 1960, the American ichthyologist Leonard P. SCHULTZ revised the genus Symphysodon and set up two new subspecies of S. aequifasciatus: S. a. axelrodi and S. a. haraldi, so that automatically the nominal subspecies S. a. aequifasciatus had to be set up, too (article 47a of the International Nomenclature rules): The result is the existence of three subspecies.

Another subspecies was established in 1981 by the American ichthyologist Warren E. BURGESS who introduced S. d. willischwartzi as subspecies of S. discus; accordingly, the nominal subspecies S. discus discus had to be set up.

The Swedish scientist Sven O. KULLANDER examined specimens of S. aequifasciatus and published his results in the 1986 work "Cichlid fishes of the Amazon River drainage of Peru". About SCHULTZ's subspecies he wrote: "I do not find any significant differences.- I cannot repeat SCHULTZ's counts.- The squamation is doubtful...". Consequently, KULLANDER classified both subspecies as synonyms of S. aequifasciatus. The same is probably true for S. d. willischwartzi (of course, as a synonym of S. discus, not of S. aequifasciatus), but in this book, KULLANDER only shortly mentions the fish, maybe because S. discus is endemic to the Rio Negro.

„Are there one or two discus species?" This question is put forward by aquarists again and again, but, until today, no scientist could answer this question once and for all. Unfortunately, some breeders have attempted to interbreed specimens of S. discus and S. aequifasciatus. They even succeeded but the beauty of neither parent could be traced in the offspring. That interbreeding of two different species is indeed possible, has already been proved in other, closely related species and is no valid proof that the two described, valid species should be united in one single taxon (please see p. 22 for further remarks on discus systematology). Sometimes a hybridisation can have a positive result; to most aquarists, this is not the case in the cross-bred discus with the single, dark, uniform middle band.

The distribution area of the discus reaches from the complete central Amazon River basin of Brazil to the Rio Putumayo in Peru. Specimens exported every now and then from the Rio Nanai are the results of deliberate or accidental releases by humans and cannot be regarded as proof of a natural distribution (at least, I was told so in Iquitos and Heiko BLEHER agreed with that point of view). As there are several tributaries of the Amazon River which have not been exhaustively examined (for example, the Rio Japurá that joins the Amazon at the town of Tefé) we can expect some imports of rare or unknown discus strains in the future.

Let´s start with listing the well-known habitats near the Amazon mouth. The waters of the lower Amazon are the distribution area of the so-called Brown Discus, a variant of S. aequifasciatus. Their distribution ranges approximately up the town of Santarém and the oppositely placed town of Alenquer (GÜNTHER, 1862; REGAN, 1905; HASEMAN, 1911; MAYLAND: ›Diskusfieber‹, 1988, p. 60f.).

S. discus is known to occur in the Rio Trombetas which joins the Amazon further west and comes from the northern highlands of Guyana. In the area between Alenquer and the Trombetas/Amazon joint, one can find populations of different variants, like, for example, forms looking very much like the 'Royal Blue'. From the south, the Amazon is joined by the clearwater streams Rio Tocatins, Rio Xingú, and Rio Tapajlenquer. These waters obviously do not meet the discus needs regarding water chemistry, because they only occur in the parts where the clearwater is already mixed with water from the Amazon (MAYLAND, 1988, STEINDACHNER, 1875).

At this point, I would like to make some remarks on the form known in trade as "Alenquer". The little town has come to some fame in the hobby because in its surrounding waters Brown Discus with a high amount of red were found. These fish very soon were highly popular in the hobby, but in fact animals with much red had only been sorted out from large catches of the Brown Discus. Very soon, the collectors began to sort out all specimens with a considerable amount of red, that had been collected in the "neighborhood" and called them "Alenquer". Now, if you think, that all Brown Discus that display the nice red colouration come from the waters near the town of Alenquer, you think wrong. Or put

quer‹ müßten aus der engeren Umgebung dieses Ortes am Amazonas stammen, der irrt! Man sieht: Den Namen ›Alenquer‹ muß man eher als Handelsmarke sehen. Dabei ist es wichtiger, sich nach der Qualität der zu erwerbenden Tiere zu richten als nach ihrer namentlichen Herkunft! Kein Fänger und kein Exporteur verrät die genaue Lage der Fangplätze von Tieren, die ihm viel Geld einbringen. Ich konnte einige wenige nur mit dem Versprechen, keine Information weiterzugeben, bereisen — und ich möchte gern noch einmal wiederkommen!

Ströme, die zur Kategorie der Weißwasserflüsse gerechnet werden, müssen nicht zwangsläufig diskusfeindlich sein, wenngleich ihr Wasser wegen der mitgeführten Sedimente fast undurchsichtig ist. In diese Ströme münden hin und wieder Schwarzwasserflüsse ein, und wo sich Weiß- und Schwarzwasser mischen, können *Symphysodon* durchaus existieren. Das beste Beispiel dafür bieten die Gewässer um die Stadt Maués sowie andererseits der Rio Abacaxís, dessen Bett etwa 140 km weiter westlich parallel zum Rio Madeira verläuft. Aus dem Rio Abacaxís beschrieb BURGESS 1981 die Unterart *Symphysodon discus willischwartzi* und damit die einzigen bisher bekannten Populationen des sogenannten Heckel-Diskus (*S. discus*) südlich des Amazonas (vgl. Foto in MAYLAND: ›Diskusfische — Könige Amazoniens‹, 2. Aufl., Seite 113), die wohl einwandfrei der Art *S. discus* zugerechnet werden können. Dagegen kommen neuerdings Importe (Aquarium Glaser, Importeur Ottlik und andere) mit der Herkunftsbezeichnung ›Rio Maués-Açú‹ aus dem gleichnamigen Fluß nahe der Stadt Maués. Sie haben eine rötlichbraune Färbung, erinnern im Habitus an den braunen Diskus der Art *S. aequifasciatus* (vgl. DATZ-Sonderheft ›Diskus‹, 1996, Seite 60), doch fehlt diesen Tieren jegliche Musterung auf den Kiemendeckeln, wofür sie eine kräftigere schwarze mittlere Querbinde auszeichnet. Ob in diesem Zwischenbereich eine natürliche Hybridisierung stattgefunden hat, wird erst noch durch genaue Untersuchungen festzustellen sein. Tiere ähnlichen Typs werden auch unter der Herkunftsbezeichnung ›Río Içá‹ in den frühen 70er Jahre mit einer (!) Expedition von Heiko BLEHER eingeführt und an den damals noch sehr aktiven Dr. SCHMIDT-FOCKE in Bad Homburg weitergegeben. Der Rio Içá (im Mittel- und Oberlauf ›Río Putumayo‹ genannt) ist jedoch vom Rio Maués-Açú mehr als 1.200 Kilometer entfernt, weshalb man beide Namen nicht verwechseln sollte.

Der Río Negro mit einigen seiner Nebenflüsse ist das eigentliche Verbreitungsgebiet des Heckel-Diskus (*S. discus*), wie nicht allein Johann NATTERER bereits vor mehr als 150 Jahren feststellte, sondern auch viele Autoren

another way: To aquarists, the (trading) name of a fish is not as important as its actual look.

Streams that belong to the so-called "white water rivers" do not have to be discus-hostile, although the sediments these rivers carry along make the water almost non-transparent. Every now and then, such a white water river is joined by a black water stream, and the resulting mixture of both water types are an environment where Symphysodon *do occur. The best example for this are the waters surrounding the town of Maués, the drainages of the Rio Madeira, which consist of the mentioned water mixture and which are famous for the only known population of the Heckel Discus <u>south</u> of the Amazon. These fish that were described in 1981 by* BURGESS *as the subspecies S .d. willischwartzi (please compare the photo in* MAYLAND*: "Diskusfische - Könige Amazoniens", p.113) are definitely S. discus, so one has to doubt that today's imports (Aquarium Glaser, Importer Ottlik [both Germany], and other importers) allegedly coming from "Rio Maués-Açú" have been collected in the same biotopes (please compare DATZ special issue "Diskus", 1996, P. 60). They have a reddish-brown colouration and are very similar to the Brown Discus of the species S. aequifasciatus. But they are lacking any markings on the operculum and have a broad dark band like S. discus. Further investigastions will show if in this area a natural hybridization among the two species of discus happened. In the early seventies, from one (!) expedition discus of a similar look were imported by Heiko* BLEHER *under the name ›Río Içá‹. He gave the fish to Dr.* SCHMIDT-FOCKE *in Bad Homburg, Germany, who was a very active and prominent breeder in those days. The Rio Içá (whose middle and upper parts are called "Rio Putumayo") is more than 1200 kilometres away from Maués!*

The main distribution area of the Heckel Discus (S. discus) is the Rio Negro and some of its tributaries. This has been reported by Johann NATTERER *150 years ago and many authors of our century (*REGAN *1905;* HOLLY, MEINKEN & RACHOW *1927;* MAYLAND *1981, 1988, 1995;* GEISLER*, 1972, 1986 and others).*

Here, the fish is distributed from the mouth to past the point where the Rio Branco flows into the Rio Negro. Lately it was announced that the distribution ranges from the Rio Xeriunini and the Rio Jufaris to the Rio

dieses Jahrhunderts (REGAN 1905; HOLLY, MEINKEN & RACHOW 1927; MAYLAND1981, 1988, 1995; GEISLER, 1972, 1986a und andere). Seine Gewässer reichen praktisch vom Mündungsbereich bis hinter den Zufluß des Río Branco. Wie wir heute wissen, reicht die Verbreitung hier über den Río Xeriunini und den Río Jufaris hinaus bis zu Río Demini. Ich selbst konnte mich davon überzeugen, daß im östlich von Manaus gelegenen Río Preto da Eva noch Tiere von S, discus vorkommen, wohingegen im östlich anschließenden Río Urubú S. discus von S. aequifasciatus abgelöst wird. Die in letzter Zeit von einigen Aquarianern und Züchtern für ›besonders interessant‹ empfundenen sogenannten Blaukopf-Heckel werden bevorzugt aus dem Río Jauaperi eingeführt, der südlich des Río Branco in den Río Negro mündet. Bei einem Besuch Brasiliens im Jahre 1993 konnte ich mich mit meinem Freund und Exporteur S. CHAVES MONTEIRO selber davon überzeugen.

Der Rio Purús und der etwa gegenüber einmündende Rio Manacapurú sind für ihre besonders schönen Diskus-Formen (›Royal Blue‹) des Typs ›Blauer Diskus‹ bekannt. Bei der Form ›Royal Blue‹ handelt es sich um Tiere einer jeweiligen Wildfang-Auslese, <u>nicht</u> dagegen um eine Zuchtform! Diese Tiere kommen bis in die Oberläufe beider Flüsse vor und auch in den diesen Flüssen angeschlossenen Seen. Im Falle des Rio Purús ist das besonders beachtenswert, denn dieser Fluß ist nicht nur besonders windungsreich, sondern auch von enormer Länge, denn er entspringt in Bolivien! Da die früher bekannten Lebensräume im Unterlauf und später auch im Mittellauf halbwegs leer gefischt wurden, sind die Fangkolonnen inzwischen bis ins 1.540 Kilometer von Manaus entfernte Lábrea vorgestoßen. Man kann sich vorstellen, welch lange Fahrten die kleinen Fangboote mit den Diskusfischen von diesem Platz aus bis zur Sammelstelle Manaus zu bewältigen haben! Ein ›Zuckerschlecken‹ kann das für die Fische somit keinesfalls sein.

Im Gebiet um den Rio Coari fängt die Region an, aus der die Tiere mit dem Farbschlag ›grün‹ stammen, also diejenigen, die früher als die Unterart *Symphysodon aequifasciatus aequifasciatus* geführt wurden. PELLEGRINS Exemplare, die seiner Beschreibung zugrunde lagen, waren aus Tefés Gewässern entnommen. Tefé liegt am Fluß gleichen Namens, der zum Einzug des oberen Amazonas (den man auch Rio Salimões nennt) gehört, genau wie der Rio Coari. Züchter, die sich besonders der Vermehrung der rotgepunkteten Tiere zugewandt haben, finden in den ausgesuchten (!) Exemplaren aus dem Río Coari und dem Río Tefé dazu das beste Basis-Zuchtmaterial, das dann mit Tieren anderer Herkunft gekreuzt

Demini. I could personally see that S. discus occurs in the Rio Preto da Eva which is east of Manaus, whereas further east, in the Rio Urubú, S. discus is replaced by S. aequifasciatus. Lately, specimens of the Blaukopf-Heckel that are preferred by some hobbyists have been collected mainly in the Rio Jauaperi which flows south of the Rio Branco into the Rio Negro. I visited Brazil in 1993 and together with my friend, the exporter S. CHAVES MONTEIRO, I saw it with my own eyes.

The Rio Purús and the approximately opposite joining Rio Manacapurú are especially known for the occurrence of the "Royal Blue" discus. The variant "Royal Blue" is always a selected wild caught, never a commercially developed breeding-form! The fish are endemic to the upper parts of both rivers and the connected lakes. In the case of the Rio Purús this fact is most interesting because this rivers is not only very winding, it is also very long: it has its rise in Bolivia! As the former collecting localities in the lower and middle part of the river have fished more or less "discus-free", the collectors have now reached collecting sites near the town of Lábrea, as far as 1540 kilometres away from Manaus. One can imagine how long it takes the small fishing boats to return with their catch to the collecting point in Manaus! It is certainly no holiday trip for the fish!

The area around the Rio Coari is known to be the home of the "green form" of S. aequifasciatus, i.e. the fish that had been classified as the subspecies S. aequifasciatus aequifasciatus because PELLEGRIN had collected the fish he used for his description near Tefé. The town Tefé is placed at the river with the same name, that belongs to the drainages of the upper amazon (that is also called Rio Salimões), just like the Rio Coari.

Breeders who have concentrated on the reproduction of the red-spotted animals find the best material for their purpose in selected (!) specimens from the Rio Coari and the Rio Tefé. This "material" is then interbred with specimens of a different orgin and then, through severals generations, developed to hgh-quality forms like "Red Spotted Green" (LO WING YAT, Taiwan).

There is no need to say much about the Tefé-Discus, but, just like in humans, one has to say that each single animal is more or less beautiful; there are, indeed, differen-

und in Hochzucht über viele Generationen zu Formen wie ›Red Spotted Green‹ weiterentwickelt wurde (LO WING YAT, Taiwan). Über den Tefé-Diskus viel Worte zu verlieren, sollte eigentlich überflüssig sein, doch gilt erstens auch hier der Grundsatz, daß (wie bei uns Menschen) nicht jedes Exemplar die gleiche Schönheit hat und zweitens ein Unterschied in Musterung und Färbung zwischen den Tieren aus dem vom nahen Amazones beeinflußten Lago Tefé und dem dunkles Schwarzwasser führenden Rio Tefé besteht, wobei die letzten zu den aquaristisch begehrteren (weil mit höherem Grünanteil versehenen) zu rechnen sind.

In der letzten Zeit haben sich einige wenige Exporteure damit befaßt, Diskusfische in den anschließenden Gewässern des Weißwasser führenden Rio Japurá (mündet gegenüber von Tefé) zu beschaffen und zum Versand zu bringen. Auch diese Tiere gehören der Kategorie ›grün‹ an. Sie sind inzwischen bereits in die Zuchtprogramme einiger Züchter integriert worden. Diskusfische, die oberhalb der Rio bzw. Lago Tefé-Einmündung gefangen und aus Manaus angeboten werden, sind selten. Das hängt vor allem mit den Riesenentfernungen in Amazonien zusammen und verbunden mit der Tatsache, daß die Fänger rechnen müssen. Tefé ist von Manaus rund 660 Kilometer entfernt, und bei einer Fahrgeschwindigkeit der kleinen, meist mit älteren Dieselmotoren ausgerüsteten Fangbooten von kaum mehr als 8 bis 10 Stundenkilometern, sind An- und Rückfahrt schon lange genug. Diskus-Angebote, die aus einem Gebiet westlich von Tefé stammen (wie etwa Tiere aus dem Rio Içá, dessen Mündung von Manaus fast doppelt so weit, nämlich 1.285 Kilometer entfernt ist), können normalerweise bestenfalls aus Leticia in Kolumbien stammen und dann über die Hauptstadt Bogotá exportiert worden sein. Derzeitige ›Erzählungen‹ von Rio-Içá-Importen müssen (vorerst) Legenden bleiben. Die einzigen bekanntgewordenen bisher eingeführten Tiere sind — wie erwähnt — das Fangergebnis einer Expedition Anfang der 70er Jahre, die BLEHER seinerzeit unternahm und die Tiere Dr. SCHMIDT-FOCKE zuführte. Importe aus Lima und Iquitos in Peru kommen dagegen aus dem westlichsten Diskus-Verbreitungsgebiet, dem Rio Putomayo und den von Menschenhand geschaffenen Revieren im Rio Nanai nahe Iquitos. Die Zukunft und damit die Entwicklung der brasilianischen Inland-Preise, deren Währung inzwischen dem US-Dollar gleichgesetzt ist, wird uns lehren, was und wieviel woher kommt.

ces between fish that come from the Amazon influenced Lago Tefé and the black water river Rio Tefé, the fish from the latter being acknowledged to be the more beautiful (because having more green in their colouration).

Lately, some few exporters have managed to collect discus in waters connected to the white water river Rio Japurá (that flows into the Amazon just opposite of Tefé) and to market them. These fish, too, belong to the category 'green', and they are already integrated into commercial breeding programmes.

Discus that are collected above the Rio/Lago Tefé confluence and are offered from Manaus are really rare. This is mainly due to the fact that all distances in the Amazon region are enormous: Manaus and Tefé are 660 km away from each other and the local collectors travel with old motorboats with weak engines that are no faster than 8-10 km/H - one can see immediately that this distance alone is incredibly long. Discus that are offered from regions west of Tefé (like specimens from Rio Içá that is twice as far away from Manaus, viz. 1285 km) can only be collected at Leticia in Columbia and then exported via the capital Bogota this distance alone is incredibly long.

For the time being, present "tales" of imports from the Rio Içá have to be considered "tall" ones. The only speciemens that were definitely imported alive were the result of an expedition in the early 70´s, that were led by Heiko BLEHER who gave the animals to Dr. SCHMIDT-FOCKE.

Imports from Lima and Iquitos in Peru, on the other hand, do come from the most western distribution area, the Rio Putomayo and the man-made habitats in the Rio Nanai near Iquitos. The future and thus, the to be expected national prices in Brazil (whose currency is equal in value with the US dollar) will tell us how many and where from the fish are imported to Germany.

Diskuszucht in und Diskuszüchtungen aus Südostasien

Die Diskuszüchter Südostasiens sind nicht nur fleißige, sondern auch erfinderische Menschen. Wer sich einmal die ›Zuchthäuser‹ vieler Malaysier (vornehmlich auf der Insel Penang), Thailänder und Chinesen (in Singapur, Hongkong wie in Taiwan) angesehen hat, muß begreifen, daß die meisten unserer ›Nach-Feierabend-Züchter‹ ihnen gegenüber nur Aquarianer sind, die dieses Geschäft in meist sehr bescheidenem Umfang betreiben. Das hat in erster Linie mit der sozialen Struktur dieser Länder, andererseits aber auch mit den ›gottgeschaffenen‹ tropischen Verhältnissen zu tun. Als ich mich vor ein paar Jahren im Rahmen der Aquarama in Singapur mit einigen Züchterfreunden, die vorher nur wenig mit europäischen Züchtern zu tun hatten, traf und ganz beiläufig nach dem pH-Wert ihres Leitungswassers fragte, erhielt ich zur Antwort: „pH-Wert? Was ist das?" Das hat nichts mit »Andere Länder, andere Sitten« zu tun: Diese Leute wurden nie vor die Frage nach dem Funktionieren eines Zuchtwassers gestellt! Es ist einfach da! Natürlich gibt es auch andere, belesenere und gelehrtere. Der pH-Wert in diesen Zuchtwässern ist nicht überall unbedingt ideal, und oft liegt er weit über der Neutralgrenze von 7,0 (GEISLER, 1996b, S. 62). Diese Alkalität hat auch in diesen Ländern mit dem Leitungsschutz zu tun, denn ein saures Wasser mit aggressiver Kohlensäure würde mit der Zeit das Rohrnetz schädigen. Darum wird auch dort das Wasser vor dem Transport durch das Netz der Leitungsrohre auf einen Wert im alkalischen Bereich eingestellt. Durch die starke Bevölkerungsdichte in einigen dieser Regionen und den keinesfalls damit einhergehenden öffentlichen wasseraufbereitenden Maßnahmen kann auch der Nitratwert des Wassers nicht an die Idealwerte Amazoniens heranreichen. Trotzdem: Die Wasser-Gesamthärte bzw. die elektrische Leitfähigkeit liegen im extrem niedrigen Bereich (nach GEISLER [1996b] im Durchschnitt bei 1,5 °d bzw. 70–80 µS). Ein regelmäßiger Teilwasserwechsel ist Pflicht, aber die Wasser- bzw. Abwasserpreise sind in diesen Ländern nicht mit den europäischen Maßstäben vergleichbar und die Wärme liefert die Sonne ebenfalls frei Haus. Viele Züchter verfügen in ihren Häusern über eine Entnahmestelle für Grundwasser. Zudem braucht man für die aufgestellten Zuchtbecken keine abgeschlossenen Räume, sondern stellt sie zum Beispiel auf selbstgebaute Metall- oder Holzgestelle in einer Gartenanlage, in der nur die jeweilige Aquarienreihe mit einem schattenspendenden und regenabhaltenden Dach versehen ist. Ich habe auch Züchter kennengelernt, die ihr Wasser vor der züchterischen Verwendung entkeimt

Commercial discus breeding in Southeast Asia

The commercial breeders of SE Asia are not only very busy but also very clever people. Whoever has once seen one of the commercial 'farms' in Malaysia (mainly on the island of Penang), Thailand or China (in Singapore, Hongkong and Taiwan), knows that, compared to these huge "plantations", our "leisure time breeders" are hobbyists who breed fishes only in minor quantities. The reasons for this obvious superiority can be found in the social structures of these countries, but also in the existing natural, tropical conditions.

A few years ago, I talked at the Singaporean Aquarama to some breeders who are friends of mine and never really dealt with European breeders. In the course of the chat I asked them at which pH they bred their fishes, only to get the astonished reply: "pH value? What's that?" This answer had nothing to do with the fact that different countries have different customs; they simply never had been confronted with the need to provide 'correct' breeding water! It is simply there! Of course, there are also other, more educated breeders. The pH value of the water of these regions is not always optimal, very often it is way above the neutral value of 7.0 (GEISLER, 1996b, p. 62). This alkalinity of the water is due to the pipeline protection we know from our countries. Acidic water containing aggressive carbonic acid would, sooner or later, damage the pipelines; this is why the water is adjusted to an alkaline level before it is pumped into the pipeline system. Above that, often these regions are quite crowded with people and the treatment of the water is mostly not in accordance with the population density. Thus, especially the nitrate concentration is not as optimal as in the natural habitat, the Amazon. Still, the total hardness and conductivity values are extremely low (with an average of 1.5° d and 70-80 µS (GEISLER [1996b]). A regular, partial water change is absolutely necessary, but the prices for water and sewage reprocessing cannot be compared to European standards. they are much lower. Therefore, the breeders can easily afford to set up such huge breeding farms: many breeders have their own ground water wells.

Also, they do not have any electricity costs because the sun is always shining, providing warmth and light. In these regions, there is no need for closed halls for the breeding tanks; they are simply set up in the backyard, only covered by a roof sheltering from too much sun or

oder anderswie vorbereitet haben. Dazu gehört das Entfernen des Chlors, der aus entkeimenden Gründen dem Wasser zugefügt wird wie auch ein Absenken des zu hohen pH-Wertes (man sieht, es gibt doch Züchter, die sich mit den pH-Werten auskennen!). Meistens trifft man bei den Züchtereien auf große Kunststoff-Vorratstanks, in denen das aufbereitete Wasser für die kommende Verwendung belüftet und in hohen Literzahlen gespeichert wird.

Zur Vermehrung der Diskus in Asien wurde bereits an anderer Stelle (vgl. MAYLAND: ›Diskusfische — Könige Amazoniens‹, 2. Aufl. 1992, Seite 176) berichtet. Hier sollen nur die wichtigsten Daten nochmals zusammengestellt werden. So wird beobachtet, und das haben auch züchterische Kontrollen in Europa und den USA ergeben, daß die Laichabgabe der Diskus-Weibchen durch aufkommenden Tiefdruck (der hier zuweilen das Einsetzen der amazonischen Regenzeit simulieren kann) positiv beeinflußt wird. Das Männchen gibt dazu Signale durch vibrierendes Flossenschlagen in der hinteren Körperhälfte und seine Schwanzflosse färbt sich rußig. Es sind die Signale der Natur, die wohl überall auf der Welt gelten und ihren Einfluß auch fern der amazonischen Heimat der Fische spürbar werden lassen. Die Regenzeit in Südostasien fällt dort in unsere Herbst/Winter-Saison, also etwa von Oktober bis Februar. Während dieser Zeit verzeichnen die Züchter ihre größten Vermehrungserfolge mit Diskusfischen. Erfolg spornt an, und so ist es nicht allein die hohe Nachkommenzahl, die den südostasiatischen Berufszüchtern Geld in die Kasse bringen soll: Es sind vor allem auch die ständig erweiterten Zuchtformen der Diskusfische, mit der uns unsere tropischen Diskusfreunde überraschen.

Sich in der Szenerie der asiatischen Zuchtformen auszukennen, ist nicht einfach, denn es werden alle paar Monate Tiere mit neuen Farben und Musterungen vorgestellt und vor allem — einer asiatischen Tradition folgend, wie wir sie von der Koi-, Goldfisch- und der Zucht Lebendgebärender Zahnkarpfen kennen — mit neuen, oft phantastisch klingenden Namen belegt. Erinnert sei hier nur an eine der derzeit erfolgreichsten Zuchtformen mit den schönen Namen ›Red Dragon‹ oder ›Pigeon Blood‹, wobei von letzterem viele Aquarienfreunde bis heute noch nicht wissen, daß er in der Übersetzung soviel wie ›Taubenblut‹ bedeutet, was sich auf einen gleichnamigen Edelstein bezieht. So schwirren viele Namen englischen Ursprungs durch die Diskus-Szene, deren Träger in der Werbung auch auf Fotos abgebildet sind. Nur kann man dabei nicht sagen, ob es sich dabei

rain. I have met several breeders who conditioned the water they used for breeding tanks in one or the other way, like, for example, sterilising it. For that, they first add chlorine for sterilisation, than remove it; further treatment includes lowering the too high pH value (which proves that there are indeed some Asian breeders who know a lot about pH!). These breeders mostly store the conditioned water in large plastic tanks where the water is aired while waiting for further usage.

Reproduction has been reported about in great detail in other publications (compare, for example, MAYLAND: Diskusfische - Könige Amazoniens, 2. Auflage, S. 176). In the USA and Europe, checks have been carried out, proving that the females' readiness to spawn increases with low pressure (which simulates in our regions the beginning of the rainy season in the Amazon.) The male reacts with vibrating fins in the posterior body half and a change of colour of the caudal fin which turns sooty black. The signals of nature are recognised everywhere by the fish, whether they are in Southeast Asia or Germany or Amazonia. In the SE Asia, the rainy season falls on our autumn/winter season, i.e. from October to February. During this time of the year, the breeders produce the highest rates of discus offspring.

Such success encourages the breeders to not only produce many, but also the most beautiful discus, so that our SE Asian friends regularly surprise us with new, breathtaking breeding forms of the fish. It is not easy to find one's way around the irritating multitude of Asian breeding forms of the discus. Fish with 'new' colours and body markings are introduced every few months and - following an old Asian tradition - named with the most fantastic terms, like the at the moment most popular forms "Red Dragon" and "Pidgeon Blood". Right now, many fish with such bizarre English names are offered, but one never knows whether these fish are reproducible strains or simply single, 'accidentally' emerged specimens which are sold at incredible prices. "Turquoise Discus", "Platinum Gold Discus", "Ocean Green Discus", "Sunshine Discus", "Blue Diamond Discus", "Snakeskin Discus", "Striated-Red Discus", "Red Alencer" (pure nonsense!), "Hi Body Blue Discus", "Great Blue Discus", "Hi Blue Discus", "Vertical Discus", "Golden Dragon Discus", "Solid Pastel Green Discus", "Blue Head Ghost", "Red Spotted Green Discus", "Red Marlboro Discus", "Tangerine Discus", etc., etc. Do you known them all?

um Zuchtstämme oder einzelne, aus dem Zuchtprogramm ›herausgefallene‹ Exemplare handelt, die dann zu horrenden Preisen und gegebenenfalls mit entsprechenden Abnahmeverpflichtungen angeboten werden.

›Turqoise Discus‹, ›Platinum Gold Discus‹, ›Ocean Green Discus‹, ›Sunshine Discus‹, ›Blue Diamond Discus‹, ›Snakeskin Discus‹, ›Striated-Red Discus‹, ›Red Alencer‹ (welch ein Unsinn!), ›Hi Body Blue Discus‹, ›Great Blue Discus‹, ›Hi Blue Discus‹, ›Vertical Discus‹, ›Golden Dragon Discus‹, ›Solid Pastel Green Discus‹, ›Blue Head Ghost‹, ›Red Spotted Green Discus‹, ›Red Marlboro Discus‹, ›Tangerine Discus‹,usw. usw. Kennen Sie sich aus (hoffentlich spätestens nach der Lektüre des Bildteils dieses Bandes)? Wie einfach wirkten da die ursprünglichen ›deutschen‹ Namen wie ›Flächigtürkis‹, ›Rottürkis‹, ›Brillanttürkis‹, ›Perlrot‹ oder selbst ›Royal Blue‹! Aber in einem Zeitalter, in dem viele unserer Jugendlichen englische Hits mitsingen, ohne selber der englischen Sprache mächtig zu sein, darf man sich über solch verwirrenden und manchmal schnell wechselnden Angebote nicht wundern. Man soll nicht glauben, daß ich, wo ich mich derart auslasse, etwas gegen die Züchterkünste der Asiaten einzuwenden hätte! Es wird aber keine Jahrzehnte mehr dauern (vorausgesetzt der Boom hält an), da werden wir mit den ersten Diskus-Gen-Manipulationen aus Fernost konfrontiert. Wetten?

Skalare

Seitdem der deutsche Dr. H. LICHTENSTEIN im Jahre 1823 mit *Zeus scalaris* den ersten Segelflosser wissenschaftlich beschrieben hatte, waren die frühen Aquarianer Ende des 19. Jahrhunderts schon sehr gespannt auf den ersten Import der Fische mit der außergewöhnlichen Gestalt und Beflossung. So wurden denn auch die ersten, im Jahre 1909 durch Siggelkow (Hamburg) eingeführten Skalare wie Kostbarkeiten behandelt und blieben jahrzehntelang das aquaristische Aushängeschild engagierter Aquarianer und mit der Aquaristik verbundener Firmen.

Die Gattung *Zeus* war bereits 1758 von LINNAEUS aufgestellt worden (*Zeus faber* = ›Heringskönig‹) und geht wiederum auf eine Beschreibung von ARTEDI zurück. *Zeus* ist eine Gattung für Meeresfischen, die in eine andere Ordnung (Zeomorphi) gehört, wogegen der Skalar zu den Barschartigen, den Percomorphi, zählt. CUVIER überführte die Art *scalaris* in die von ihm selbst 1816 aufgestellte Gattung *Platax*, deren Gattungstyp *P. teira* FORSKÅL, 1775 ist. Wahrscheinlich sah CUVIER deshalb eine

How is one supposed to keep track with all these (at least to us Germans and other Europeans) foreign names and terms? Don't get me wrong in saying this, but the tempo with which new breeding forms are developed leads towards more and more confusion and developments like the new genetic engineering will definitely even increase the speed. I think, it won't take long - if the discus doesn't loose its popularity - until the first, Asian bred, genetic manipulated specimens are marketed!

Angels

In 1823 the German scientist Dr. H. LICHTENSTEIN described the first angel Zeus scalaris. This fish has been one of the most popular aquarium fish ever since and when, at the end of the 19th century more and more people became involved in the hobby, they - of course - were all eagerly waiting for the first import of the notorious fish with the strange body shape and remarkable finnage. In 1909, Siggelkow (Hamburg/Germany) imported the first specimens; they were treated like precious jewellery and for a very long time, the angel was an absolute "must" for every advanced hobbyists and an advertisement for aquarium fish trading companies.

The genus Zeus had been set up by LINNAEUS in 1759 (Zeus faber) who again referred to a description by ARTEDI. But Zeus is a genus of marine fishes, so that CUVIER placed in 1816 the species scalaris in the genus Platax which he had himself had set up. The genus type is P. teira FORSKAL, 1775. I think, CUVIER probably placed

enge Verwandtschaft, weil *P. teira* (ebenfalls ein Meeresbewohner) ein ebensolcher Hochflosser wie der Skalar ist. Erst HECKEL rückte mit seiner Beschreibung von *Pterophyllum* die ›Sache‹ ins rechte Licht, weil er erkannte, daß die Art *scalaris* aufgrund der Gestalt ihrer Schlundknochen zu den Chromiden (den heutigen Cichliden) gehört, die man jedoch damals noch zu den Lippfischen, den Labroiden zählte, wie man aus dem Text im folgenden Abschnitt erfahren kann. MÜLLER trennte später (1844) die Chromiden von den Labroiden ab und 1889 schließlich verwendete COPE anstelle des bis dahin gebräuchlichen Familiennamens Chromidae den neuen Namen Cichlidae.

In seiner Gattungsbeschreibung für *Pterophyllum* merkt HECKEL zu dem vorhandenen Material, das LICHTENSTEIN hinterlegt hatte, folgendes an: »Die Verfasser haben den Fisch, von dem sie nur ein verstümmeltes Exemplar in der Bloch'schen Sammlung zu Berlin unter dem Namen *Zeus scalaris* fanden, einstweilen in ihre Gattung *Platax* gestellt.Sein äusseres Ansehen kommt allerdings jenem, der zu dieser Gattung gezählten Fischen, sehr überein, allein die Gestalt seiner Schlundknochen gestattet keinen Zweifel über die Stelle, welche wir ihm hier anweisen; auch sind weder *Platax* noch sonst ein Squamipenne (wenn man *Scatophagus* und *Toxotes* davon ausnimmt) Bewohner süsser Gewässer. Diese *Chaetodon*-Gestalten in den Flüssen Brasiliens gehören durch ihre Schlundknochen ebenso den Labroiden, wie jene in den Flüssen Indiens durch ihre labyrinthförmigen Organe den Labyrinthiformes an.« Man erkennt aus diesen Sätzen die relative Unerfahrenheit mit all den neuen Geschöpfen, die NATTERER mit seiner Sammlung präsentierte. Am Ende seiner genauen Beschreibung stellte der Autor HECKEL fest: »Dieser seltene Fisch, von welchem das k. k. Museum acht Individuen besitzt, wurde in einem Igarapé (Waldbache) in der Nähe von Barra do Rio Negro mit dem Wurfnetze gefangen, er wird nicht über 5 Zoll lang.«

Es hat in den zurückliegenden Jahren und nach HECKELS Aufstellung der Gattung *Pterophyllum* mit *P. scalaris* als Gattungstyp eine Reihe Beschreibungen gegeben, wie die von *Plataxoïdes dumerilii* CASTELNAU, 1855 und *Pterophyllum eimekei* AHL, 1928. Dazu kommen weitere Beschreibungen dieses Jahrhunderts, wie *Pterophyllum altum* PELLEGRIN, 1903 und *Plataxoides leopoldi* GOSSE, 1963, der inzwischen ebenfalls der Gattung *Pterophyllum* angeschlossen wurde.

Viele wußten mit all den Namen nicht viel anzufangen,

the angel in Platax *because* P. teira *is also a high-finned fish. It was* HECKEL *who placed the fish in the genus* Pterophyllum *after he had described it because he had discovered that the species scalaris had to be assigned to the group of Chromids (today: Cichlids. As you will see in the following, the Chromids were then regarded as Labroids.), due to the form of their gullet bones. In 1844,* MÜLLER *separated the Chromids from the Labroids and finally, in 1889,* COPE *introduced the family name Cichlidae instead of Chromidae - the name, these fishes still have today.*

In his description of the genus Pterophyllum, HECKEL *made some remarks on the material that had been passed on by* LICHTENSTEIN: "Die Verfasser haben den Fisch, von dem sie nur ein verstümmeltes Exemplar in der Bloch'schen Sammlung in Berlin unter dem Namen Zeus scalaris fanden, einstweilen in ihre Gattung Platax gestellt. Sein äusseres Ansehen kommt allerdings jenem, der zu dieser Gattung gezählten Fischen, sehr überein, allein die Gestalt seiner Schlundknochen gestattet keinen Zweifel über die Stelle, welche wir ihm hier anweisen; auch sind weder* Platax *noch sonst ein Squamipenne (wenn man* Scatophagus *und* Toxotes *davon ausnimmt) Bewohner süsser Gewässer. Diese* Chaetodon-*Gestalten in den Flüssen Brasiliens gehören durch ihre Schlundknochen ebenso den Labroiden, wie jene in den Flüssen Indiens durch ihre labyrinthförmigen Organe den Labyrinthiformes an."[1] *One can gather from these sentences that the scientists were really inexperienced with the fishes* NATTERER *presented in his collection. At the end of his detailed description,* HECKEL *summarises: "Dieser seltene Fisch, von welchem das k. k. Museum acht Individuen besitzt, wurde in einem Igarapé (Waldbache) in der Nähe von Barra do Rio Negro mit dem Wurfnetze gefangen, er wird nicht über 5 Zoll lang."[2]*

---

1 *"The authors have temporarily placed the fish, of which they have found only one, mutilated specimen in the Bloch Collection /Berlin under the name Zeus scalaris, in the genus Platax. Its outer appearance indeed resembles very much the fishes of this genus but the form of its gullet bones leaves no doubt about the place we put this fish in; also, neither Platax nor any other Squamipenne (except from Scatophagus and Toxotes) live in fresh water. These Chaetodon-figures from the rivers of Brazil with their gullet bones belong to the Labroids just like the ones of the rivers of India with their labyrinth-like organs belong to the Labyrinthiformes."*

2 *"This rare fish, of which the k. k. Museum possesses eight specimens, was captured in an Igarapé (forest stream) with a net, it does not grow larger than 5 inches."*

und nur einige Verwegene benutzten sie mit der bescheidenen Kenntnis, die ihnen in aquaristischen Zeitschriften vermittelt wurde. Selbst Hermann MEINKEN, der allseits beliebte und in der Zeit nach dem zweiten Weltkrieg anerkannte und von VDA bestellte Leiter der Fischbestimmungsstelle stellte *Pterophyllum eimekei* noch als ›Eimeke's Zwergsegelflosser‹ mit der Herkunftsbezeichnung »Mittellauf des Amazonas von Santarém aufwärts bis etwa Villa Bella, sowie dessen Nebenflüsse Tapajoz, Trombetas, Jamunda usw.« vor. L.P. SCHULTZ, der US-amerikanische Ichthyologe, sah 1967 *P. dumerilii* noch als gültige Art an und verwies *P. leopoldi* in die Synonymität. Dem widersprach KULLANDER 1986 deutlich, stellte charakteristische Unterschiede fest, sah SCHULTZ' Angaben als ›nicht nachprüfbar‹ an und endete mit dem Satz: »Hence I refer *P. dumerilii* to *P. scalare* and consider *P. leopoldi* a valid species« (»Hiermit verweise ich *P. dumerilii* [als ein Synonym] zu *P. scalare* und betrachte *P. leopoldi* als gültige Art«).

Pterophyllum altum PELLEGRIN ist vielen Aquarianern als ›Hoher Skalar‹ ein Begriff. Ich (MAYLAND) berichtete bereits 1976 (Aqu. Mag.) über den Fang der Tiere wie auch des Roten Neon in ostkolumbianischen Gewässern. Im selben Jahr aber veröffentlichte BURGESS (T.F.H.) eine Arbeit über Segelflosser des Rio Negro (!), verwies auf die meristischen Unterschiede beider Arten (*P. altunm* und *P. scalare*), sah dabei die unterschiedlich langen Flossen als geographische Variante zwischen Rio Negro- und Orinoko-Arten an — und synonymisierte *P. altum* mit *P. scalare*!! Diese Folgerung in der Beurteilung basierte auf SCHULTZ' meristischen Daten aus dessen Arbeit von 1967. In einer späteren Arbeit (1979) nahm BURGESS eine Teilkorrektur vor und erkannte die Orinoko-Bewohner als eine nördliche (Orinoko-)Variante und damit als Unterart von *P. scalare* an. Erst KULLANDER (1986) festigte erneut die Meinung, die jedem Aquarianer, der die Fische seither gepflegt hatte, klar sein mußte: *Pterophyllum altum* ist eine ›gute‹, eine gültige Art!

Die Verbreitung von *P. scalare* überspannt ein weites Gebiet. Sie reicht im Bett des Amazonas von Peru (Yarina Cocha/Rio Ucayali [EIGENMANN & ALLEN, 1942, KULLANDER, 1986, LÜLING, 1961] , unterer Rio Napo [KULLANDER, 1986], Rio Salimões/Tabatinga, Rio Tefé, Rio Coari, Rio Manacapurú; Rio Negro, Rio Trombetas, Óbidos, Santarém, Rio Xingú [STEINDACHNER, 1875, PELLEGRIN, 1903, REGAN, 1905, HASEMAN, 1911]) bis an die Mündungsregion nahe Belém (Rio Tocantins, Ilha de Marajó [REGAN, 1905]). Funde aus Nebenflüssen kenne ich vom Rio Madeira, dem Rio Araguaia, dem mittleren und unteren Rio

After HECKEL's description of Pterophyllum *with genus type* P. scalaris, *the past has seen several other descriptions, like* Plataxoides dumerilii *CASTELNAU, 1855 and* Pterophyllum eimekei *AHL, 1928. Further, there are the descriptions of* Pterophyllum altum *PELLEGRIN, 1903 and* Plataxoides leopoldi *GOSSE, 1963 which is now also assigned to the genus* Pterophyllum.

Many hobbyists did not know how to handle these names and only few, bold aquarists dared to use them, after having consulted the available specialist literature. Even Hermann MEINKEN, the popular and acknowledged expert who was appointed by the VDA as head of the Fischbestimmungsstelle introduced Pterophyllum eimekei *as "Eimeke's Zwergsegelflosser", coming from the central part of the Amazon, from Santarém upwards to Villa Bella, as well as the tributaries Tapajoz, Trombetas, Jamunda etc. L. P. SCHULTZ, the American ichthyologist, regarded in 1967* P. dumerilii *a valid species and valued* P. leopoldi *as a synonym. This classification was clearly denied by KULLANDER in 1986; he identified characteristic differences, rejected SCHULTZ's statements as 'not verifiable' and finished off with the sentence: "Hence I refer* P. dumerilii *to* P. scalare *and consider* P. leopoldi *a valid species."*

To many hobbyists, Pterophyllum altum *PELLEGRIN is also known as the 'Longfinned Angel'. I (MAYLAND) reported as early as 1976 in an aquarium magazine about the catching of these animals, together with the Cardinal tetra in east Colombian waters. In the same year, BURGESS published an article for T.F.H. publications about angels from the Rio Negro (!). He compared the meristic differences between both species (*P. altum *and* P. scalare*) and considered the differing lengths of the fins as geographical variants of Rio Negro and Orinoco species. And, as a result of this classification, he synonymized* P. altum *and* P. scalare*!! This result was based on SCHULTZ's meristic data from his 1967 paper. Several years later, in 1979, he partly corrected himself and recognised the Orinoco-endemic fish as a local (Orinoco-) variant, and thus classified it as a subspecies of* P. scalare. *But it was not until 1986, when KULLANDER published his paper, that the overall view of hobbyists who had indulged themselves in the care of* Pterophyllum *was confirmed by a scientist:* Pterophyllum altum *is a 'good', a valid species!*

Negro und dem Rio Branco. Aus dem Rio Negro berichtet AXELROD 1976 (Igarapé Anapichi und I. Apania). Ein weiteres Verbreitungsgebiet schließt sich im Nordosten Südamerikas an: In den Guayana-Ländern, deren Flüsse nicht über den Amazonas, sondern direkt in den Atlantik geleitet werden. Hier gibt es eine Verbreitung vom Essequibo (LADIGES, 1951) und Demerara River bis zum Rio Oyapock an der nördlichen Grenze von Brasiliens Bundsstaat Amapá.

Die Verbreitung von *Pterophyllum leopoldi* beschränkt sich — soweit wir bis heute wissen (GOSSE, 1963) — auf einen relativ eng begrenzten Raum im Bett des Rio Solimões, wie der Amazonas in seinem Lauf von der kolumbianischen Grenze bis zur Mündung des Rio Negro genannt wird. Hier erstrecken sich seine Lebensräume etwa von der Stadt Manacapurú etwa 90 Kilometer stromaufwärts.

Pterophyllum altum kennen wir nur vom mittleren Rio Orinoco, der hier einen Teil der Grenze zwischen Kolumbien und Venezuela bildet und auch in Nebenflüssen beider Länder vorkommt. Ich fand die Tiere im Rio Inirida, dem wahrscheinlich größten Fangplatz für Exporttiere aus Kolumbien, die von Puerto Inirida mit dem Flugzeug nach Bogotá gesandt und von hier in alle Welt exportiert werden.

Parallelentwicklungen (Konvergenzen), wie man sie bei vielen Cichliden auf allen Kontinenten ihrer Verbreitung kennt, gibt es auch bei Skalaren: Im Meerwasser kennen wir Hochflosser zum Beispiel bei *Platax teira* (›Langflossen-Fledermausfisch‹) aus der Familie der Spatenfische oder Ephippididae und aus dem Brackwasser die Flossenblätter (Monodactylidae), als deren langflossigster Vertreter *Monodactylus sebae* angesehen werden muß. Direkte verwandtschaftliche Beziehungen lassen sich zwischen den genannten Arten und den Skalaren allerdings nicht ausmachen.

Es wurde in den letzten Jahren viel über das unterschiedliche Stirnprofil bei Skalaren gerätselt. Der einwärts gerichtete Knick vor den Augen ist nicht allein als Erkennungsmerkmal von *P. altum* zu werten. Es gibt auch Importe aus peruanischen Gewässern, deren Exemplare einen solchen Knick aufweisen. Sie aber deshalb ›Peru-Altum‹ zu nennen ist nicht nur unsinnig, sondern verfälscht für weniger Eingeweihte das Bild des *Pterophyllum*-Komplexes. Es besteht wohl kein Zweifel darüber, daß es bei der weiten Verbreitung von *P. scalare* verschiedene lokale bzw. geographische Formen

The distribution of P. scalare *is widely spread. It includes the Amazon basin from Peru (Yarina Cocha/Rio Ucayali [EIGENMANN & ALLEN, 1942, KULLANDER, 1986, LÜLING, 1961], lower Rio Napo [KULLANDER, 1986], Rio Salimões/Tabatinga, Rio Tefé, Rio Coari, Rio Manacapurú; Rio Negro, Rio Trombetas, Óbidos, Santarém, Rio Xingú [STEINDACHNER, 1875, PELLEGRIN, 1903, REGAN, 1905, HASEMAN, 1911])) up to the Amazon mouth near Belém. Some other localities are the Rio Madeira, the Rio Araguaia, the central and lower Rio Negro and the Rio Branco. AXELROD (1976) reports from the Rio Negro (Igarapé Anapichi and I. Apania). Another distribution area follows in the north-east of South America: in the Guyana countries whose rivers do not join the Amazon but flow directly into the Atlantic Ocean. Here, the fish can be found from Essequibo (LADIGES, 1951) and the Demerara river to the Rio Oyapock at the northern border of Amapá, one of Brazil's federal states.*

The distribution of Pterophyllum leopoldi *is limited, as far as we know today (GOSSE, 1963), to a relatively restricted area in the Rio Solimões basin; Rio Solimões is the name of the part of the Amazon that reaches from the Colombian border to the Rio Negro confluence. Here, P. leopoldi is distributed over an area of about 90 km upstream from the town of Manacapurú.*

Pterophyllum altum *is only known from the central Rio Orinoco which is partly the border of Columbia and Venezuela and whose drainages stretch over both countries. I collected the species in the Rio Inirida which is probably the largest collecting site in Columbia. The fish are brought via aeroplane from Puerto Inirida to Bogotá and from there, they are exported into the world.*

Parallel developments (convergence) which are known from many other cichlid species, also occur in angels: In the ocean, long-finned species like the Platax teira, *the Longfinned Batfish of the family of Ephippididae can be found, and from brackish water, the Monodactylidae with* Monodactylus sebae, *the Finger Fish, are notorious for their long fins. Of course, this resemblance of appearance does not hint towards a close relationship between the listed fishes and angels.*

In the past years, experts have given the differently shaped profiles of angels much thought. The inward bend

gebildet haben, deren Vertreter nicht nur über eine leicht geänderte Streifenmusterung, sondern auch über eine unterschiedlich lange Beflossung verfügen. Viele Züchter, die sich bevorzugt mit der Entwicklung neuer Zuchtformen beschäftigt haben, wußten diesen Umstand zu nutzen.

In Bezug auf die **Lebensbedingungen** stellen Skalare ähnliche Ansprüche wie die Diskusfische, mit denen sie über weite Gebiete auch zusammen leben. Das läßt sich leicht an dem Umstand feststellen, daß in jedem Diskus-Netzzug stets auch auch eine stattliche Anzahl von Skalaren erbeutet werden konnten. In einigen solcher Gebiete ist die Kopf- bzw. Stirnpartie mit länglichen blutroten vertikalen Flecken übersät, die eine ähnliche Ursache haben können, wie die roten Tüpfel auf den Körpern bestimmter Diskusfisch-Populationen.

Wie die Diskusfische konnten auch die Skalare, soweit bisher bekannt, im Wesentlichen nicht über das Mündungsgebiet der drei großen, von Süden kommend in den Amazonas mündenden Klarwasserflüsse (Rio Tapajós, Rio Xingú und Rio Tocantins) eindringen. Berichtet wurde von Vorkommen am unteren Rio Xingú nahe der Mündung bei Porto do Moz, am unteren Rio Tapajós nahe Santarém sowie am unteren Rio Tocantins und auf der Ilha de Marajó im Mündungsdelta (STEINDACHNER, 1875, REGAN, 1905 und andere). Es liegen auch Belege aus dem Rio Araguaia vor, der zu System des mittleren bzw. oberen Tocantins gerechnet wird. Die Verbreitung im Bereich des Essequibo River wurde ausführlich in LADIGES' bekanntem Werk ›Der Fisch in der Landschaft‹ (1954) beschrieben, wobei der Tatsache besondere Bedeutung zukommt, daß P. scalare hier nicht, wie sonst üblich im zwischen Ästen und Unterwassergesträuch in Ufernähe, sondern (ebenfalls in Ufernähe) zwischen Pflanzen und Röhricht, aber auch an und über Klippen angetroffen wurden. Der Autor berichtet: »Bei Fangversuchen flüchteten die Pterophyllum nämlich genau wie die anderen Cichliden oft seitwärts liegend sofort in die engen Spalten des Gesteins und ließen sich nur unter größten Schwierigkeiten an den Flossen herausziehen«.

Skalare ernähren sich, ähnlich wie Diskusfische, fast ausschließlich von fleischlicher Kost. Obgleich ihre Maulspalte nur unwesentlich tiefer eingeschnitten ist als die der Diskusfische, sind sie nach meiner Erkenntnis eher bereit, Jungfische und Süßwassergarnelen einer bestimmten Größe aufzunehmen als Diskusfische. Je nach Lebensraum, so wird vermutet, nehmen auch Skalare bevorzugt Mückenlarven der Gattungen Chirono-

at the eyes cannot be regarded a mere characteristic of P. altum. There are imports from Peruvian waters with specimens that have the same king of profile. If one would call them - accordingly - 'Peru-Altum', one would not only be simply wrong but also falsify, especially for the less informed hobbyists, the whole Pterophyllum complex. There is no doubt that, due to the wide distribution of P. scalare, some local or geographical variants have developed, specimens of which display a different stripe pattern and differing finnage. Many breeders who were especially involved with the breeding of new forms knew how to use this fact for their own purposes.

*In regard of **living conditions**, angels have nearly the same requirements as the discus with which they coexist in many habitats: every netting of discus brings up a large number of angels, too. Specimens of some of these habitats have their head or forehead covered with elongated, vertical, blood red spots which could have the same origin as the red spots in certain discus populations.*

Like the discus, the angel could not get past the confluences of three big clear water rivers that flow from the south into the Amazon. There are reports of populations in the lower Rio Xingú near the mouth at Porto do Moz, the Lower Rio Tapajead or foreherém as well as in the lower Rio Tocatins and on the Ilha de Marajthree big clear water rivers that flow from the south into the Amazon. There are reports of populations in the lower Rio Xingú near the mouth at Porto do Moz, the Lower Rio Tapajos near Santarém as well as the lower Rio Tocantins and on the Ilha de Marajó in the mouth delta (STEINDACHNER, 1875, REGAN, 1905 and others). There is also roof of angels occuring in the Rio Araguaia that belongs to the system of the central and upper Tocantins. The distribution in the Essequibo River has been reported on in great detail in LADIGES' well-known book "Der Fisch in der Landschaft" (1954). Here, it is explicitly mentioned that in this habitat, Pterophyllum scalare does not (like it usually does) hide between tangled roots and submerged plants near the river banks, but can be found (also near the banks) between plants and reeds and among rocks. The author reports that when the men tried to catch them, the angels, like other cichlids, dashed for cover, often flat on the side, and squeezed themselves into cracks in rocks so that they could

mus und *Culex* auf, doch gibt es besonders die Stechmückenlarven der Gattungen *Nematocera* und *Culex* nicht überall. Bei ihnen stechen nur die Weibchen, denn sie benötigen das gesaugte Blut für die Eientwicklung. Stechmücken- bzw. Moskitolarven hängen gattungstypisch mit dem Ende des Hinterleibes unter der Wasseroberfläche und atmen durch ein kurzes Rohr atmosphärische Luft. Sie fehlen vor allem in den sehr nährstoffarmen Schwarzwässern, wie wir sie zum Beispiel vom Rio Negro kennen. Anders dagegen die Chironomidenlarven der Zuckmücken, die sind mit vielen Arten vertreten sind. Sie leben bevorzugt im Boden, und es gibt weltweit kaum eine Ansammlung von Süß- oder Brackwasser, in der man sie nicht antrifft, soweit nur die benötigten Nährstoffe vorhanden sind. Es handelt sich bei ihnen nicht um Stechmücken! Aquarianer kennen sie als ›Rote Mückenlarven‹, doch gibt es auch Formen anderer Färbung.

Zur Pflege von Skalaren benötigt man höhere Aquarien als allgemein üblich. Das gilt besonders für die hochflossigen Tiere von *Pterophyllum altum*, für deren Pflege ein Becken schon eine Höhe ab 65 Zentimeter haben sollte. Als Begleitfische empfehlen sich nur Tiere ruhigerer Arten, zur Vermehrung der Fische sollte man ein Artaquarium ohne Begleitfische vorziehen, denn auch diese — darunter auch Welse — können den Fortpflanzungsablauf stören oder gar die meist auf ein *Echinodorus*-Blatt abgegebenen Eier fressen. Ebenso warnen muß man vor einer Gesellschaft mit Barben und bestimmten Salmlern oder Lebendgebärenden, die den Skalaren die schönen langen Filamente ihrer Flossen abfressen. Von letzteren werden bestimmte Arten jedoch gern deshalb zu den Skalaren gesetzt, weil die Jungfische der Zahnkarpfen von den Skalaren als begehrte ›Appetithappen‹ angesehen werden. Auch wenn fast alle Segelflosser aus extrem weichen Gewässern stammen, müssen ihre halbwegs domestizierten Nachkommen deshalb keinesfalls in so weichem Wasser gepflegt werden. Ich habe Skalare jahrzehntelang erfolgreich in Frankfurter Leitungswasser (17 °dGH und pH 7,3) gehalten und dabei keine Probleme feststellen können. Um sie rationell zu vermehren, sollten die Wasserwerte allerdings unter den genannten liegen. Wasserflöhe, Kleinkrebse und Mückenlarven bilden auch im Aquarium die von den Fischen bevorzugte Nahrung. Bei zu einseitiger Fütterung (auch mit zuviel Flocken- oder Granulatfutter) kann es nach wiederholter eigener Erkenntnis (aber auch der von Züchterfreunden) irgendwann zur Aufnahmeverweigerung von Nahrung kommen.

only be removed with great difficulty by being pulled out by their fins.

Similar to discus, angels are almost exclusively carnivorous. As their mouth is larger than the discus', they can take in young fish and shrimps of a certain size much easier than their discus-relatives. In certain habitats, one assumes, the angels prefer to eat midge larvae of the genera Chironomus *and* Culex, *but especially the mosquito genera* Culex *and* Nematocera *do not occur everywhere. In these genera, only the females bite, because they need the blood for egg-development. Mosquito larvae hang genus-typically with their rear ends beneath the water surface and breath air with the help of a short tube. They do not occur in the nutrient-deficient blackwater regions, like the Rio Negro, unlike the* Chironomus *midges of which many species live in the ground. There is almost no brackish or fresh water patch where these animals do not occur, as long as there are enough nutrients. But these are no biting insects! Aquarists know them under the name 'blood worm' which hints at the red colour of the larvae; in the tropics, also other forms with different colours can be found.*

Keeping angels requires aquaria that are higher than the usual ones. This is especially true for aquaria with the longfinned Pterophyllum altum: *these tanks should be at least 65 cm high. As tank comrades one can only recommend quiet species, if you want to breed them, you have to keep them in a species tank, because all co-inhabitants (even catfish) can disturb the reproduction sequence or eat the eggs which are mostly laid on a leaf of* Echinodorus. *It is just as dangerous to mix angels with barbs or certain tetras or livebearers because these fishes like to gnaw at the angels' filamentous finrays. The latter are preferably commuted with angels because the young of toothcarps are very much enjoyed as 'appetisers' by the angels. Although* Pterophyllum *are endemic to extremely soft waters, there is no need to keep its offspring in water that is just as soft. I have kept angels successfully for many years in tap water of my home town (17°dGH and pH 7,3) and never had any problems. In order to reproduce them rationally, the water parameters should be below the ones mentioned above. In the aquarium, angels prefer water-fleas, brine shrimp nauplii and mosquito larvae. If you feed them too unbalanced (e.g. too much flake food), they will, sooner or later, refuse to take any food.*

Ideale Verhältnisse vorausgesetzt, ist die **Fortpflanzung** der Segelflosser nicht sehr problematisch. Die Paarbildung geschieht am besten derart, daß man aus einem gemeinsam aufgezogenen Jungfischschwarm zwei Tiere aussucht, bei denen sich bereits eine gegenseitige Sympathie erkennen läßt. Sie wird dann auch im Zuchtaquarium andauern und ›Früchte‹ tragen. Bei den typischen Offenbrütern lassen sich die Geschlechter am besten auf diese Weise finden, weil sie anhand äußerer Merkmale kaum oder nur schwer erkannt werden können. Als bevorzugter Laichplatz wird, wie erwähnt, ein Pflanzenblatt gewählt, zuweilen können es aber auch Holz und Steine sein. Das Substrat wird kritisch begutachtet und peinlichst gesäubert. Das Gelege wird von Weibchen Ei für Ei nebeneinander abgelegt und sofort vom Partner befruchtet. Große Gelege können 800 bis 1.000 Eier umfassen. Wie Diskusfische bilden die Alttiere eine sogenannte Elternfamilie, bei der sich beide an der Laich- und Brutpflege beteiligen. Die abhängig von der Wassertemperatur nach etwa 2 Tagen geschlüpften Larven werden umgebettet und an einem neuen Platz angeheftet, wo sie sich dank eines von den Kopfdrüsen gebildeten Sekrets festhalten können.mit ihrem Kopfdrüsensekret am neuen Platz angeheftet. Nach weiteren 4 bis 5 Tagen schwimmen die Jungfische frei und können schon bald nach dem Anfüttern in ein separates Aufzuchtbecken überführt werden.

Im Verlauf der seit Jahrzehnten bekannten Nachzuchten wurde eine Vielzahl an interessanten **Zuchtformen** herausgebildet. Bekannte und immer wieder begehrte Formen gehören inzwischen zum festen Bestandteil des Angebotes im aquaristischen Handel. Als ständig verfügbare Zuchtformen kann man die folgenden nennen:

Marmorskalar (›Marbleized Angel‹), Schwarzer oder Rauchskalar (›Black Angel‹, ›Black Lace Angel‹), Schleierskalar (›Veiltail Angel‹), Schwarzweißer Skalar (›Black-White Angel‹), Falbskalar (›Dun Angel‹), Leopardskalar (›Leopard Angel‹), Zebraskalar (›Zebra Angel‹), Goldener Skalar (›Golden Angel‹), Blaßroter Skalar (›Pink Angel‹), Geisterskalar (›White Ghost Angel‹)

Neben diesen seit langem eingeführten Zuchtformen tauchen gelegentlich solche Formen auf, die man zunächst einmal als züchterische Zufallsprodukte ansehen muß, bevor eine derartige Linie als so gefestigt ansehen kann, daß dieser Form ein dauernder Platz auf dem aquaristischen Markt zugebilligt werden kann.

If they have ideal conditions, the breeding of angels is no problem at all. Pairs should be chosen from a swarm of young fish, by picking two fish that already display a certain affection towards each other. This affection will last in the breeding tank and, after some time, 'bear fruits'! Angels are typical substratum-spawners, and the sexes can only be told apart through intense observation of the spawning parents, because there are no external differences between the two sexes. The preferred spawning site is, like I have already mentioned, a plant leaf, but sometimes also a stone or a piece of wood is chosen. After the substrate has been carefully selected it is cleaned thoroughly. The female lays the eggs one by one, tightly to each other, on the leaf where they are immediately fertilised by the male. Large spawns can contain 800 to 1000 eggs! Like discus, the fish form so-called parent-families in which both parents take care of the brood and the young. The larvae hatch (depending on the water temperature) after about 2 days. The parents immediately move the fry to another place where they are attached with a secretion from the head. After another 4 to 5 days the young swim free and soon after they have taken the first food they can be removed to a separate rearing tank.

In the course of many years of angel breeding, many interesting breeding forms have been developed. Today, the aquarium trade offers many popular and well-known forms; some of the continuously marketed breeding forms are:

Marbleized Angel, Black Angel or Black Lace Angel, Veiltail Angel, Black-White Angel, Dun Angel, Leopard Angel, Zebra Angel, Golden Angel, Pink Angel, White Ghost Angel.

Beside these forms that have been known in the hobby for quite some time now, rare breeding forms are marketed every now and then. These specimens should be regarded as mere coincidental products of commercial breeding efforts, until a reproducible strain has been developed and is regularly offered in the shops.

Zur Artenfrage bei Diskus und Skalaren - Anmerkungen aus biologischer Sicht

von Frank Schäfer

Die Frage, was eine Tierart ist, beschäftigt die Biologen, seit man erkannte, daß es Evolution gibt: daß Arten also nicht aus einem einmaligen göttlichen Schöpfungsakt entstanden und von da ab unveränderlich sind, sondern sich kontinuierlich aus anderen, austerbenden Formen entwickeln.

Das früheste, gerade in der Ichthyologie (der Fischkunde) bis heute angewandte Artmodell ist das der "künstlichen Art". Das bedeutet, als Art wird definiert, was sich in irgendwelchen zähl- oder messbaren Daten, in anatomischen Feinstrukturen oder der Färbung signifikant von anderen, ähnlichen Formen abgrenzen läßt. Mit diesem Modell kommt man aber in vielen Bereichen nicht weit: viele, schon rein intuitiv als unterschiedliche Arten aufzufassende Formen würden demnach zur gleichen Art gehören, auf der anderen Seite variieren selbst innerhalb einer Population (also einer natürlichen Fortpflanzungsgemeinschaft) viele klassische zählbare Merkmale (Schuppenzahlen, Flossenstrahlen, Indices etc.). Anders ausgedrückt: Es fällt unheimlich schwer, verbindlich klarzumachen, warum dieses oder jenes Merkmal so typisch sein soll, daß es geeignet ist, Arten voneinander abzugrenzen. Je nach individueller Neigung der Wissenschaftler gibt es daher "Splitters", die bereit sind, jedes noch so kleine Detail als artrelevant anzusehen und dementsprechend viele Arten anerkennen und "Lumpers", die es bei unsicheren Kandidaten vorziehen, lieber von einer variablen Art zu sprechen und nur wenige, gut definierte Arten anerkennen. Man sieht also, daß diese Form der Artabgrenzung in vielen Fällen ein künstliches, den natürlichen Verhältnissen nicht entsprechendes Bild liefert.

Neuere Ansätze versuchen daher, in der Artabgrenzung neben der erdgeschichtlichen Entwicklung und ihren Einflüssen auf die Tiere die natürlichen Verwandtschaftsverhältnisse der zu untersuchenden Formen untereinander zu berücksichtigen. Das klingt jedoch einfacher, als es ist. Erst in den letzten 70 Jahren wurde es aufgrund der rasanten Entwicklung der Aquaristik überhaupt möglich, Beobachtungen an lebenden Tieren in die Artdiagnose einzubeziehen. Etwa zwei Drittel der bis heute bekannt gewordenen Fischarten wurden aber aufgrund von einigen wenigen konservierten Exemplaren bestimmt. Die Zuordnung lebender Tiere zu diesen Originalbeschreibungen ist häufig ein nur schwer zu lösendes Problem. Hinzu kommt noch, daß überhaupt nur kleiner Prozentsatz aller auf der Erde lebenden Arten jemals in Gefangenschaft gehalten wurde.
Fast alle Systematiker sind sich heute darüber einig, daß die Grundeinheit einer jeden Tierart die Population ist. Also Tiere, die untereinander eine natürliche Fortpflanzungsgemeinschaft bilden und ihr genetisches Material permanent untereinander austauschen. Sie sind miteinander uneingeschränkt fruchtbar kreuzbar. Das gleiche gilt für ihre Nachkommen in unbegrenzter Folge. Schwierig wird es aber in dem Moment, wo es gilt, die verwandtschaftlichen Verhältnisse der einzelnen Populationen untereinander zu bewerten, also die Art als sol-

About the species classification of discus and angels - from a biological point of view

by Frank Schäfer

The question, how a species can be defined, has been on biologists' minds ever since evolution is known to be a part of nature: that species did not come into being in a single, say divine, act of creation, and have stayed the way they are ever since, but that they continuously develop from older, vanishing species.

The earliest, especially in ichthyology (= the branch of zoology that deals with fishes) used species model is the one of the "artificial species". This means that a species is defined after any countable or measurable data, anatomic structures or characteristic colouration that distinguish it significantly from other, similar forms. Unfortunately, in many areas one doesn't get too far with this model: many, almost intuitively recognised, independent species would then belong to the same one, whereas even within one population (within the same, natural interbreeding community) many of the characteristics listed above differ considerably (e.g. number of scales, fin rays etc.). Put another way: It is incredibly difficult to explain reliably why certain distinctive marks should be especially suited for distinguishing the species from each other. Depending on the individually preferred scientific doctrine, there are scientist who belong to the so-called "splitters" and are ready to accept even the most tiny, differing detail to classify a species - and who, accordingly, set up a large number of valid species. And there are, on the other hand, the "lumpers" who prefer to speak of a species variety if they are not absolutely sure and who accept only few, valid species. It is easy to see that this model of species classification leads - in many cases - to an artificial representation that is not in accordance with the actual status quo in nature.

This is the reason why the latest approaches of species classification try to take into consideration not only the geological developments and their influence on the animals but also the natural relationships between the species in question. Now, this sounds far more easy than it really is. Only during the last 70 years when the hobby raised increasing interest the observation of live fish became a major issue and was included in species classification. But - about two thirds of all known species were classified in the past with few, conserved specimens. Assigning live specimens to these original descriptions is often very difficult. Adding to these problems is the fact that only a small fraction of all living species has ever lived in captivity.

Today, almost all systematists agree that the popula-tion is the basic unit of every species. Like I said before: a population is a community of animals that interbreed and thus permanently exchange their genetic material. They can reproduce with each other without any restrictions; this is also true for all offspring and all genera-tions. The really hard part is to tell different populations apart from each other and value them, i.e. to actually classify a species as such. In

che zu definieren. Im vorstehenden Text hat H. J. MAYLAND ausgeführt, daß Diskusfische und Skalare ein riesiges Verbreitungsgebiet haben. Ganz sicher kommt ein Skalar aus dem Unterlauf des Amazonas in Brasilien unter natürlichen Bedingungen nie in die Verlegenheit, sich mit einem Skalar aus dem Oberlauf des Amazonas in Peru paaren zu müssen. Aber sind es deshalb schon unterschiedliche Arten?

Um diese Frage zu erörtern, muß man sich überlegen, was die Faktoren sind, die im Laufe der Evolution dazu führen, daß sich überhaupt unterschiedliche Arten aus einer gemeinsamen Stammform entwickeln. Zunächst einmal, das wissen wir, ist die Natur ausgesprochen konservativ. Bewährte Modelle werden ungeachtet des engeren Verwandtschaftsgrades immer beibehalten. So besitzen z.B. alle Wirbeltiere rotes Blut als Transportmedium für gelöste Stoffe aller Art innerhalb des Körpers. Das ist so, weil jeder lebende Organismus auf diesem Planeten ein ungeheuer komplexes System darstellt, dessen Funktionieren einzig und allein auf einer fehlerfreien Funktion der universellen Matrix des Lebens beruht: der DNA (Abkürzung für Desoxyribonukleinsäure). Unsere eigene Existenz beweist anschaulich, daß sich dieses Molekül im großen und ganzen bewährt hat. Jedoch ist nichts auf dieser Welt fehlerfrei. Und auch an der DNA treten von Zeit zu Zeit Fehler auf. In sehr vielen Fällen sind die davon betroffenen Individuen nicht lebensfähig. In anderen Fällen passiert überhaupt nichts, weil der Fehler auf der DNA an einer Stelle auftrat, die keine Funktion hat. In wiederum anderen, seltenen Fällen ergibt es sich, daß das von dem Fehler seiner DNA betroffene Individuum lebensfähig ist; es ist jedoch in irgendeiner Art und Weise von seinen Artgenossen unterschieden: eine Mutation (Veränderung) hat stattgefunden.

In der Aquaristik kennen wir viele solcher Mutationen. Meist beziehen sie sich auf das äußere Erscheinungsbild: pechschwarze oder reinweiße Fische, solche mit einem Goldglanz auf dem Körper oder mit stark vergrößerten Flossen; Mopsköpfigkeit oder Glotzaugen; eine verkürzte oder verkrümmte Wirbelsäule; oder auch Krebs. Diese Aufzählung ließe sich noch lange fortsetzen. Andere Mutationen fallen nicht so sehr ins Auge, obwohl sie eine wichtige Voraussetzung für die kontinuierliche Zucht der Fische im Aquarium darstellen. Innerhalb der unabdingbaren Umweltparameter, die jede Fischart benötigt, liegt ein relativ großer Spielraum. Jeder weiß, daß Diskusbuntbarsche zur erfolgreichen Zucht weiches, saures Wasser benötigen. Innerhalb eines jeden Geleges gibt es aber einige wenige Eier, die sich auch unter nicht optimalen Bedingungen entwickeln. Die daraus schlüpfenden Diskus sind schon sehr viel flexibler, was ihre Ansprüche an das Wasser angeht. Innerhalb weniger Generationen ist es auf diese Art und Weise möglich, eine Aquarienpopulation von Diskus aufzubauen, die auch in hartem, leicht alkalischen Wasser erfolgreich züchtet. Ob diese Tiere farblich und von der Körperform her dem entsprechen, was der Züchter anstreben sollte, sei einmal dahin gestellt. Wesentlich ist in dem hier besprochenen Zusammenhang nur, daß es geht.

Nun stelle man sich einmal vor, eine Population Diskus würde, z.B. nach einem Erdbeben, in einem abgeschlossenen See isoliert. Mit der Zeit würde das Wasser in dem

the introduction, H. J. MAYLAND mentioned that discus and angelfish inhabit a huge distribution area. Of course, an angel from the lower Amazon in Brazil will, under natural conditions, never interbreed with an angel from the upper, Peruvian Amazon. But does that have to conclude that these two belong to two different species?

In order to answer this question, one has to think first about the question which factors forced animals in the course of evolution to develop different species from one basic form in the first place. Nature is, first of all, conservative - that's one thing we know for sure. Reliable models are always kept up, despite any close degree of relationship. For example, all vertebrates have red blood as a means of transportation for all kinds of dissolved substances inside the body. This conservatism exists because each living organism on this planet represents an incredibly complex system whose function depends exclusively on the perfect work of the universal matrix of life: the DNA (deoxyribonucleic acid). The existence of mankind shows vividly that this molecule has, by and large, proven to be working. Still, nothing in this world is absolutely perfect. And even in the DNA, every now and then, mistakes occur. Very often, organisms with such a defective DNA cannot live. In other cases, nothing at all happens, because the mistake occurred in a part of the molecule that has no special purpose. And in some other, rare cases it happens that the affected individual can live; but somehow it is different from all other organisms of the same species. This latter kind of permanent change is called mutation.

In the hobby, we know many mutations. Most of them are represented through changes in the fish's appearance like pitch black or completely white specimens; fish with a golden shimmer or enlarged fins; puggy heads or goggle eyes; a reduced or curved backbone; or cancer. This list could be continued endlessly. Other mutations are not as obvious, although they are important prerequisites for the continuous breeding of fishes in the aquarium. Within the indispensable environmental parameters a fish species needs for its survival lies a relatively large scope. Everybody knows that discus need for successful breeding soft, acidic water. But with-in each spawn, there are some eggs that develop even if the conditions are not perfect. The fish hatching out of these eggs are much more flexible concerning the required water. With these fish, one can build up an aquarium population of discus that can be bred in hard, alkaline water within only few generations. Whether this new population represents a fish whose body shape and colour lives up to the breeder's expectations, is a completely different matter. In regard of what is said here, the only interesting thing is the fact that such a manipulation is possible.

Now, imagine a population of discus would - after, for example, an earthquake - be separated in a hermetically isolated pond. In the course of time, the water in this pond would become harder and more alkaline. This is the same effect like the one described for the aquarium. The only difference is that in nature, this process takes a much longer time, say, about 1000 years. Seen from a geological

See immer härter und alkalischer. Es träte der gleiche Effekt ein, wie er eben für das Aquarien-Experiment beschrieben wurde, nur spielt sich das ganze über einen viel längeren Zeitraum, sagen wir einmal tausend Jahre, ab. Geologisch und aus Evolutionssicht ist das ein kurzer Zeitraum, aber auf die Diskuspopulation hätte es bereits einen tiefgreifenden Einfluß: Die allermeisten der ursprünglich isolierten Tiere wären nämlich ausgestorben. Nur die wenigen, angepassten Exemplare hätten überlebt. Die Auswahl an Geschlechtspartnern wäre sehr stark eingeschränkt gewesen, damit hätte nur ein geringer Austausch genetischen Materials stattgefunden. Letzteres ist bekanntlich der Hauptgrund für die sexuelle Vermehrung, denn durch den Austausch genetischen Materials wird in erster Linie verhindert, daß fehlerhafte DNA weitervererbt wird: Jedes Elterntier stellt 50% seines Erbgutes bei der Befruchtung zur Verfügung und bei der daraus resultierenden Neukombination der Gene werden Fehler weitgehend eliminiert. Die starke Inzucht in unserem isolierten System würde spontan auftretende Mutationen aber sehr begünstigen. Geht in einer großen Population eine Mutation schnell in dem allgemeinen Erbgut auf, so können sich in einer kleinen, isolierten Population verdeckt (rezessiv) vererbte Eigenschaften schnell etablieren. Insgesamt wächst also in unserem hypothetischen See die Anzahl von Diskusfischen mit einer sehr eigenartigen DNA schnell an. Das kann soweit gehen, daß die im See lebende Population eines Tages nicht mehr mit anderen Populationen Nachkommen zeugen kann, weil die unterschiedliche DNA bei der Neukombination der Gene nach der Befruchtung nicht mehr in der Lage ist, eine Funktionseinheit zu bilden: Eine neue Art wäre entstanden.

Nun war das ein sehr dramatisches Beispiel für eine Artentstehung. Immer sind es jedoch Umweltbedingungen, die dafür sorgen, daß zufällig an diese Umweltbedingungen besser angepasste Individuen überleben und sich mit der Zeit unterschiedliche Arten ausbilden. Ist z.B. irgendein Nahrungsmittel in einem Biotop im Überfluß vorhanden, es gibt aber im Biotop keinen Fisch, der diese Nahrungsquelle ausnutzen kann, so wäre ein Mutant, der diese Futterquelle für sich erschließt, extrem im Vorteil: Er hätte sich eine ökologische Nische erschlossen. Auch daraus entstehen mit der Zeit eigenständige Arten, denn die Nachkommen des Mutanten hätten gegenüber ihren Artgenossen immer den Vorteil einer unbegrenzten Futtermenge, wodurch diese Mutation im Laufe der Evolution begünstigt würde.

Wie soll man nun aber mit diesem Wissen im Hinterkopf die Diskus- und Skalar-Formen im Hinblick auf den Artenstatus beurteilen? Offensichtlich ist, daß es sich um äußerst erfolgreiche Arten handelt, denn sie haben ein großes Verbreitungsgebiet und sind innerhalb dieses Verbreitungsgebietes häufig und in großer Individuenzahl anzutreffen. Augenscheinlich ist, daß sich die Evolution dieser Arten im Fluß befindet, denn anders sind die vielen verschiedenen Formen nicht zu erklären. Aber sind es nun eigenständige Arten, die schon so weit voneinander fort entwickelt haben, daß eine uneingeschränkte Kreuzbarkeit nicht mehr gegeben ist, oder handelt es sich um Unterarten oder gar nur um Farbvarianten?

Um diese Frage zu erörtern, muß noch einmal kurz ausgeholt werden: Was ist eine Unterart? Wir wissen, daß

or historical view, this is really a short period, but - nevertheless - this short period would be long enough to have a considerate influence on the discus population: The majority of the originally isolated fish would be extinct. Only few, genetically adjusted specimens would have survived. As the choice of partners for reproduction is extremely limited in such an isolated group, the exchange of genetic material is just as limited. This exchange of genetic material is one of the main 'reasons' for sexual reproduction: This way, the transmission of imperfect genes is prevented. In the fertilisation process, 50% of the genetic material of each parent are combined, thus forming 100% of 'new' genes; in the process, possible genetic defects are largely eliminated. The inevitably occurring inbreeding in the isolated system we talk about at the moment would support spontaneous mutations. While in a large population, such a mutation quickly dissolves in the general genetic make-up, the hidden (recessive) transmission of mutated genes happens virtually unhindered in such a small population like the one in 'our' isolated pond. As a result, the number of discus with a 'strange' DNA rises quickly. This development can reach a stage where the fish from the isolated pond cannot interbreed any longer with discus from another population because their genes could not combine to form a functioning DNA - a new species has developed.

The example above is a really dramatic one for how a new species can develop in the course of time. But - it is an irrefutable rule that environmental circumstances sort out individuals that are - accidentally - better adjusted to these certain environmental conditions, individuals, which are, on the other hand, able to survive conditions others cannot: new species come into being. Let's take another example: In a habitat, plenty of a certain food is available to fish that cannot eat this food; a mutant, being able to eat the food the other inhabitants of the habitat cannot eat, would have an obvious advantage - the exclusive exploitation of an ecological niche. This is a second way how species can develop in the course of time, as the offspring of the mutant could use this advantage; the availability of an exclusive food source would obviously favour the survival and further reproduction of the mutated specimens in the course of evolution.

Keeping this basic information about species development in mind, how can one categorise the different discus and angelfish varieties in terms of species classification? It is obvious that all discus and angelfish species are really 'successful' ones as they inhabit large distribution areas and are numerously represented in their habitats. Further, in evolutional terms, the species are quite clearly still developing - there is no other explanation for the huge number of forms existing at the moment. The question is whether these different forms are independent species, e.g. genetically so far from one another that they cannot interbreed? Or are these forms subspecies or maybe only colour varieties?

In order to answer these questions one has to go back one step and reconsider: What is the correct definition of the term 'subspecies'? We know that nearly all species which

fast alle Arten, die ein großes Verbreitungsgebiet haben, innerhalb dieses Verbreitungsgebiet variieren. Ist diese Variation sehr augenfällig und betrifft sie innerhalb der einzelnen Population alle Individuen, so bezeichnet man diese Formen als Unterarten. Der wichtigste Beweis dafür, daß es sich um Unterarten und nicht um Arten handelt, ist immer der, daß in den Gebieten, wo zwei Unterarten aneinandergrenzen, uneingeschränkt fruchtbare Mischformen auftreten, die man dann als Intergrade-Population bezeichnet. Als Farbvarianten sind hingegen solche Individuen aufzufassen, die zwar, für sich alleine gesehen, deutlich in bestimmten Merkmalen von der Norm abweichen. Diese Abweichungen beziehen sich jedoch nicht auf die gesamte Population.

Unter diesem Aspekt betrachtet fällt es zumindest für zwei Diskus-Formen relativ leicht, eine Entscheidung bezüglich ihres Status zu treffen: Der Blaue und der Braune Diskus sind ganz offensichtlich nur Farbvarianten der selben Fortpflanzungsgemeinschaft. Daß sich für die Aquarianer für lange Zeit ein anderes Bild ergab, liegt daran, daß diese Fische je nach Blauanteil in der Färbung unterschiedlich teuer sind. Die blauen Tiere werden deshalb beim Exporteur bereits aussortiert und beim Importeur entsteht deshalb der Eindruck, Blaue und Braune Diskus kämen getrennt vor; Dem ist aber nicht so. Anders die Grünen Diskus. Ihre geografischen Verbreitungsgrenzen sind genau definiert. Man fängt z.B. bei Tefé schlicht und ergreifend weder Braune noch Blaue Diskus. Auch stabile Inter-grade-Populationen sind nicht bekannt. Mehr noch: Obwohl wir beschämend wenig über die Reinzucht von Wildpopulationen des Diskus wissen (was in unseren Aquarien schwimmt sind in aller Regel entweder Wildfänge oder Haustier-Züchtungen, die Erhaltungszucht erbreiner Wildpopulationen wird leider so gut wie nicht betrieben) scheint es so zu sein, daß alle Diskusstämme, wo Blau/Braun und Grün gekreuzt wurden, konstant auf die Einkreuzung von Fremdblut oder die Rückkreuzung auf die Elterntiere angewiesen sind. Bei reiner Inzucht solcher Stämme ist eine F3 (also die dritte Generation Geschwisterpaarung) nur unter größten Schwierigkeiten zu erzielen, eine F4 liefert schon keine vermehrungsfähigen Tiere mehr. Sollten sich diese Befunde erhärten (die derzeit bekannten Ergebnisse beruhen auf statistisch nicht hinreichend abgesicherten Zufallsbeobachtungen, die unter ganz anderen Fragestellungen gemacht wurden), so würde das bedeuten, daß der Grüne Diskus eine eigenständige, vom Braun/Blauen Diskus zu unterscheidende Art ist!

Nun könnte man einwenden, daß dieser Befund auf unzureichende Unterbringung, Inzuchtschäden aufgrund mangelhafter Auslese der Zuchttiere etc. zurückzuführen ist. Dem spricht aber entgegen, daß auf Selektionsbasis, und nicht durch Einkreuzung entstandene Zuchtformen, wie der Rot-Türkis-Diskus gegen derartige Inzuchteinflüsse unempfindlich sind. Das gleiche gilt für die einzige Wildform, die, soweit sich das recherchieren ließ, tatsächlich auf Erhaltungszuchtbasis gezüchtet wird: den Braunen Diskus "Alenquer". Auch bei dieser Form ist die F4 keinerlei Problem, ein Fertilitätsverlust nicht feststellbar.

Was nun den Heckel-Diskus angeht, so schien die Sache bis vor einigen Jahren recht einfach zu sein. Der Heckel hat eine rundere Körperform und war immer gut an sei-

are distributed over a large area vary within this distribution area. If one of these varieties is very conspicuous and applicable to all individuals of one population, this one form is called 'subspecies'. The most important proof of a species' status as subspecies is the fact that in areas where two populations of obvious varieties mix, the two forms can interbreed without restriction and produce fertile offspring. This next generation is called 'intergrade-population'. Colour-varieties, on the other hand, are individuals which differ considerably in certain characteristics from other individuals of the same population - but these differences cannot be found in the whole population.

Looked at from this point of view, it is quite simple to classify two form of the discus: The Blue and the Brown Discus are obviously only colour varieties of the same reproductive group. The fact that many hobbyists thought otherwise for a long time was caused by the different prices for fish with differing quantities of blue. Exporters sort out the blue specimens beforehand so that one could get the impression, the blue and the brown variety are found separately. But they are not. In the Green Discus, matters are different. The geographic boundaries of its distribution area are exactly defined. Near Tefé, for example, one will never catch a Blue or a Brown Discus. Stable intergrade-population are neither known. And although we know shamefully little about breeding pure wild forms (the fish we have in our tanks are usually wild caughts or private breeds, genetically pure strains of wild populations are scarcely bred), it seems as though all strains of discus in which the blue/brown and the green form were interbred, are, for further breeding, dependent on "foreign blood" or reproduction with the parent specimens. After a period of pure inbreeding, such strains become almost infertile in the F 3 generation (i.e. the third generation of reproduction with brothers and sisters), the F 4 is totally infertile. If these results should be substantiated through further research (the findings described above are based on statistically not yet proven, coincidental observations that were made while investigating completely different questions), they would prove that the Green Discus is indeed an independent species and not a colour variety of the Blue/Brown Discus!

Now, one could say that these findings resulted from careless maintenance, or from genetic defects through inbreeding and inadequate selection of the breeding specimens etc. Against that one can answer that other strains that have been produced through selection and not through interbreeding, like the Turquoise/Red Discus, are absolutely immune against inbreeding. The same is true for the only (as far as I know) wild form that is commercially bred at the moment: the Brown Discus "Alenquer". In this form, too, the F 4 is absolutely fertile.

As far as the Heckel Discus is concerned, things seemed to be quite easy - until some years ago. The Heckel Discus has a rounder shape and could be easily distin- guished by the clearly broadened middle band. Also, it inhabits other areas than S. aequifasciatus. Still: in the last couple of years, specimens from the Madeira area became known which look like aequifasciatus *but have the typical "Heckel*

nem stark verbreiterten Mittelstreifen zu erkennen. Außerdem bewohnt er ein anderes Verbreitungsgebiet als *S. aequifasciatus*. Jedoch wurden in den letzten Jahren aus dem Madeira-Gebiet Tiere bekannt, die wie *aequifasciatus* aussehen, aber den typischen Heckelbalken zeigen. Hinzu kommt noch, daß gerade die Madeira-Tiere eine ungeheuer variable Färbung haben. In der Literatur (z.B. DEGEN, 1997) wurden diese Fische bislang als Naturhybriden zwischen *S. discus* und *S. aequifasciatus* bezeichnet. Es deutet jedoch, betrachtet man die Nachkommen von Madeira-Fischen (GÖBEL, mdl. Mitt.) nichts darauf hin, daß der Heckel-Diskus bei der Entstehung dieser Form beteiligt war: Die Jungfische entsprechen in ihrem Erscheinungsbild alle *S. aequifasciatus*. Allerdings, und das ist hochinteressant, wird die Anlage zum verbreiterten Mittelbalken rezessiv, das heißt verdeckt, vererbt. Eine abschließende Beurteilung der Madeira-Tiere ist zum jetzigen Zeitpunkt nicht möglich. Es handelt sich möglicherweise um eine künstliche Population, deren Grundlage entkommene Tiere unterschiedlicher Herkunft der Braun/Blauen Form von *S. aequifasciatus* sind. Dadurch kam es, daß sich die im Erbgut von allen *S. aequifasciatus* vorhandene Anlage zur Ausbildung eines breiten "Heckel-Balkens" wieder etablieren konnte. Daraus kann man schließen, daß der Ancestor (so bezeichnet man die ausgestorbene Stammart, aus der sich die jetzt lebenden Arten entwickelt haben) dem heutigen *S. discus* ähnlicher gesehen hat, was die Ausprägung der Streifen angeht, als dem heutigen *S. aequifasciatus*. Über Kreuzungen zwischen Heckel-Diskus und anderen Diskus nur sehr wenig bekannt. Eine F1 scheint relativ problemlos erzielbar zu sein, doch hat das im Hinblick auf die Artenfrage keinerlei Aussagewert.

Ob der Blaukopf-Heckel eine Unterart oder Farbvariante von *S. discus* ist, ist völlig unklar. Die Material- und Datenlage zu diesen Fischen ist mehr als dürftig, auch Felduntersuchungen liegen kaum vor.

Alle Liebhaber und Züchter von Diskusfischen könnten der Wissenschaft helfen, endlich Klarheit in die weitgehend unverstandene Systematik dieser schönen Buntbarsche zu bringen, indem sie, wenn sie Kreuzungsexperimente vornehmen, genau über die Ergebnisse Buch führen. Die daraus hervorgehenden Fische sollten fotografiert werden, verstorbene Exemplare unbedingt in 70%igem Ethylalkohl konserviert werden. An jedem konservierten Fisch ist ein Etikett anzubringen, aus dem der genaue Ursprung hervorgeht. Das Etikett ist mit Bleistift zu beschriften, da Tinten in Alkohl gelöst werden.

Der eine oder andere Leser mag sich bei der Lektüre gefragt haben, warum zur Klärung der Artenfrage bei Diskusfischen nicht einfach die DNA untersucht wird, wenn schon alle klassischen Methoden der Artabgrenzung (also Zählwerte, Färbungsmerkmale und Verhaltensunterschiede) zu versagen scheinen und Kreuzungsexperimente im Labor am dafür notwendigen Aufwand scheitern. Nun, entsprechende Untersuchungen finden derzeit tatsächlich statt. Doch die dabei erhaltenen Ergebnisse sind aus technischen Gründen stark interpretationsbedürftig und deshalb immer umstritten. Die DNA-Analyse kann bei den momentan möglichen Methoden immer nur ein zusätzlicher Puzzlestein sein, der sich in ein Gesamtbild einfügt.

band". In addition to that, these "Madeira specimens" show an incredible colour variety. In the specialist literature (e.g. DEGEN, 1997), these specimens are classified as 'natural hybrids' of S. discus and S. aequifasciatus. Still, if one examines the offspring of Madeira fishes (GÖBEL, personal communication), there is no sign that the Heckel discus was involved in the emergence of this special form: the young fish do indeed all look like S. aequifasciatus. One point is most interesting, though: the predisposition to the broader middle band is passed on recessively. At this stage, a final conclusion concerning the Madeira fish cannot be made. It might be an 'artificial' population that developed from escaped or released specimens of the blue/brown form of S. aequifasciatus from different origins. On this basis the predisposition to display the broad 'Heckel band' (which exists in the genes of all S. aequifasciatus) could re-establish itself. From this one can conclude that the ancestor (the not longer existing base form from which the current species have developed) was - as far as the band pattern is concerned - more similar to S. discus then to today's S. aequifasciatus. There is little information about interbreeds of the Heckel discus and other discus species. The F 1 seems to be bred unproblematically, but this is not meaningful at all in the problem of species classification.

Whether the Pineapple Discus is a subspecies or colour variety of S. discus, is not clear yet. Data about these fish are more than rare and useful results from scientific examinations are hardly available.

It is up to all discus enthusiasts to help science out of this dilemma: The still unsolved question of species systematisation in these wonderful cichlids could be brought a huge step further if all hobbyists who experiment with interbreeding discus species would keep an exact account of all their results. The offspring of all these experiments should be photographed, dead animals conserved in 70% ethyl alcohol. Each conserved fish should be labelled with a sticker with the exact origin of the hybrid. Please use a pencil for labelling, ink dissolves in alcohol.

I am sure that several readers - after having read this introduction so far - have asked themselves why science does not examine the fish's DNA in order to finally classify the different discus species - obviously, all 'classic' methods like behaviour examination, measure data, coloration differences have failed and interbreeding experiments in the laboratory are way too time-consuming. Well, such examinations are indeed going on at the moment. But, for technical reasons, the results from these examination methods have to be individually interpreted which makes them the subject of never ending discussions. DNA analysis is, at the moment, nothing more than another piece of the puzzle, adding to the overall picture.

Zusammenfassend läßt stellt sich die Systematik beim Diskus nach dem derzeitigen Wissensstand also wie folgt dar:

Symphysodon HECKEL, 1840
mit zwei Arten

Symphysodon aequifasciatus PELLEGRIN, 1904
mit zwei Unterarten

S. ae. aequifasciatus PELLEGRIN, 1904 (Grüner Diskus)
S. ae. ssp. (Brauner und Blauer Diskus)

Symphysodon discus HECKEL, 1840
mit zwei Unterarten:

S. d. discus HECKEL, 1840 (Heckel-Diskus)
S. d. willischwartzi BURGESS, 1981 (Blaukopf-Heckel)

Da *S. ae. axelrodi* und *S. ae. haraldi* (Brauner bzw. Blauer Diskus) in der gleichen Arbeit von SCHULTZ (1960) beschrieben wurden ist, habe ich (Frank SCHÄFER) hier keinen Namen eingesetzt. Die Autorenschaft dieser Unterart ist auf jeden Fall SCHULTZ, 1960. Beide Namen sind einander gleichwertig. Der International Code of Zoological Nomenclature sieht für solche Fälle vor (Artikel 50 f.), daß der erste Revisor den in Zukunft gültigen Namen festzulegen hat. Da dieser Beitrag nicht als Revision der Gattung *Symphysodon* gemeint ist, möchte ich auf keinen Fall mißverstanden werden und benutze hier keinen Namen für die Unterart Brauner/Blauer Diskus. Aus dem gleichen Grund werden im Bildteil dieses Buches bei den Diskusfischen keine wissenschaftlichen Unterartnamen benutzt.

Ähnlich verworren und bislang nur unbefriedigend gelöst sieht die Sache bei den Skalaren aus. Allgemein anerkannt werden:

Pt. altum PELLERGRIN, 1903
Pt. leopoldi (GOSSE, 1963)
Pt. scalare (LICHTENSTEIN, 1823)

Pt. dumerilii (DE CASTELNEAU, 1855) wird derzeit nicht als valide anerkannt. Das liegt daran, daß KULLANDER (1986) den sehr schlecht erhaltenen Holotypen untersuchte, der keinerlei Färbung mehr aufweist und dem viele Schuppen fehlen. Als einziges Kriterium, ob die Art valide ist oder nicht, konnte daher das Kopfprofil herangezogen werden. Im Gegensatz zu den eindeutig nicht zu *Pt. scalare* gehörigen Tieren, die in der Aquaristik als *Pt. dumerilii* angesprochen werden und die ein stumpfes Stirnprofil haben, weist der Holotyp von *Pt. dumerilii* ein *Pt. scalare* entsprechendes Kopfprofil auf, mit dem typischen Knick oberhalb der Schnauze. Das bedeutet, daß entweder der "Aquarium-dumerilii" (wir nennen ihn hier "Schafskopf-Skalar") eine noch unbeschriebene Art darstellt oder der traurige Erhaltungszustand des Holotypes ein falsches Licht auf die Sache wirft.

Zu diesen vier auf den ersten (oder spätestens zweiten) Blick ganz klar unterschiedenen Arten kommen noch etliche, bei denen die Unterschiede nicht so augenfällig und die daher systematisch kaum zu bewerten sind. So werden aus Peru (wo nach KULLANDER nur die Art *Pt. scalare* lebt) mindestens drei verschiedene Phänotypen (also Tiere, die sich äußerlich unterscheiden) für die Aquaristik importiert. Der bekannteste ist der sog. "Peru-

The actual scientific systematisation of the discus is as follows:

Symphysodon HECKEL, 1840
with two species:

1. Symphysodon aequifasciatus PELLEGRIN, 1904
with two sub-species:
S.ae.aequifasciatus PELLEGRIN, 1904 *(Green Discus)*
S.ae.ssp. *(Blue and Brown Discus)*

2. Symphysodon discus HECKEL, 1840
with two sub-species:
S.d.discus HECKEL, 1840 *(Heckel Discus)*
S.d.willischwartzi BURGESS, 1981 *(Pineapple Discus)*

As S.ae.axelrodi and S.ae.haraldi (Brown and Blue Discus) were described in the same paper by SCHULTZ *(1960), the author (*SCHÄFER*) has refrained from using names. The authorship of this subspecies belongs to* SCHULTZ*, 1960. Both names are of the same scientific value. In cases like this, the International Code of Zoological Nomenclature (article 50 ff.) decides that the first reviser of a genus is supposed to establish the future valid name. As this article is not intended to be a revision of the genus* Symphysodon*, I want to protect myself from any misunderstandings and simply do not use any scientific names for the subspecies Blue/Brown Discus. Due to the same reason, the captions in this book do not use scientific names either.*

A similar confusing situation can be discovered in the angelfishes. Commonly accepted are:

Pt. altum PELLEGRIN, 1903
Pt. leopoldi (GOSSE, 1963)
Pt. scalare (LICHTENSTEIN, 1823)

Pt. dumerilii *(DE CASTELNEAU, 1855) is, at the moment, not considered a valid species because* KULLANDER *(1986) examined the badly conserved holotype which had lost all its colour and many of its scales. The only criteria for evaluation was the shape of the head. In contrast to the animals that clearly do not belong to* Pt. scalare *because of their even profile and are called 'Pt. dumerilii' in the hobby, the examined holotype of 'Pt. dumerilii' had a typical 'Pt. scalare' profile with the clearly recognisable bend above the snout. This means that either the 'aquarium-dumerilii' (in this book, we call the fish "Sheepshead Angel") is an undescribed species or that the bad shape the holotype is in shows the matter in the wrong light.*

Adding to these four, at first (or at least, second) glance clearly distinguishable species are several others; still, in these forms, the distinguishing features are hardly recognisable and, therefore, not useful for systematising the genus. For example, from Peru (where, according to KULLANDER*, only one species, Pt. scalare, occurs), at least three different phenotypes (i.e. specimens which have different appearances) are imported for the hobby. The best known is the so-called 'Peru-Altum' which, as* MAYLAND *already explained, has nothing to do with the 'real' Pt. altum. This fish stands out because of its high body and a striped cau-*

Altum", der jedoch, wie MAYLAND bereits ausführte, mit dem "echten" Pt. altum nichts zu tun hat. Dieser Fisch fällt durch seine hohe Gestalt auf und besitzt eine deutlich gestreifte Schwanzflosse. Die mittlere senkrechte Körperbinde (die oft nur verwaschen zu erkennen ist) läuft über den gesamten Körper, also vom Ansatz der Rückenflosse bis zum Ansatz der Bauchflosse. Eine sehr auffällige Variante dieses Fisches hat noch zusätzlich viele rote Tüpfelchen auf dem ganzen Körper. Die dritte Form aus Peru unterscheidet sich von anderen Skalar-Wildfängen aus Brasilien durch einen großen, bohnenförmigen Fleck hinter dem Auge und dadurch, daß die mittlere Binde nur etwa zur Körpermitte reicht. Es bestehen außerdem bei manchem Importen Unterschiede in der Streifung des weichen Teils der Rückenflosse.

Aus Surinam werden Skalare importiert, die einen deutlich scheibenförmigeren Körper haben bei gleichzeitig nicht so hohen Flossen. Auch von dieser Form gibt es eine Variante mit zahlreichen roten Punkten auf dem Körper.

Die wenigen Wildfänge aus Brasilien, die ich bisher zu Gesicht bekommen habe, entsprechen eher dem Typ "eimekei", das bedeutet, die Tiere haben eine extrem hohe Stirn, die fast rechtwinklig zum Rücken abknickt, wenn sie etwas größer sind.

Die Gattung *Pterophyllum* ist jedenfalls stark revisionsbedürftig. Leider besteht kaum Interesse seitens der Aquarianer an Wildfangtieren "normaler" Skalare. Es ist bei der Bedeutung der Fische für die Aquaristik auch oft schwer, zu entscheiden, ob eingeführte Tiere wirklich aus autochtonen (also ursprünglich dort vorhandenen) Populationen stammen oder ob es sich um Nachfahren ausgesetzter Tiere handelt. Wegen der geringen Nachfrage macht sich auch kein Exporteur die Mühe, aus weiter abgelegenen Fanggebieten Skalare mitbringen zu lassen.

Die Tatsache, daß die drei oben erwähnten Peru-Formen oft zusammen eingeführt werden, heißt nicht, daß sie auch in der Natur zusammen vorkommen. Gängige Praxis bei den Exportstationen ist, alles was die Fänger bringen, solange in einem Tank zu sammeln, bis eine Sendung sich lohnt. So erreicht uns eigentlich eher durch Zufall, nämlich weil die "Peru-altum" in der Aquaristik begehrt sind, auch Wildfangmaterial anderer Formen.

Es bleibt zu wünschen, daß sich die Aquarianer wieder verstärkt den Wildformen der Diskus und Skalare zuwenden, um mitzuhelfen, daß wir (nicht zuletzt unter dem Aspekt des Artenschutzes) endlich die Naturgeschichte dieser Fische verstehen. Man kann nur schützen was man kennt. Bislang kennen wir nur die Haustiere. Ihre wildlebenden Ahnen bleiben vorerst voller Rätsel.

dal fin. The middle vertical band (that is often quite faded) runs over the whole body, from the base of the dorsal to the base of the pelvic fin. A very conspicuous variety of this fish has additional red spots all over the body. The third form from Peru can be distinguished from the other wild Brazilian forms by a large, bean-shaped spot behind the eye and the middle band that reaches only the middle of the body. In some imports, there are also differences in the stripes on the soft part of the dorsal fin.

From Surinam, specimens are imported that have a clearly more disc-shaped body and less high fins. There is a variety with red spots of this form, too.

The few wild caughts from Brazil I have seen in my life look like the type 'eimekei' which means that the fish have an extremely high forehead that appears almost vertical to the back in adult specimens.

The genus Pterophyllum *needs a revision, indeed. Unfortunately, hobbyists seem not to be interested in wild caughts of the 'common' angelfish. But it is really difficult in this genus to decide whether the imported specimens come from original populations or if they are descendants of released animals. As the demand for these fish is so little, no importer takes the trouble to import angelfish from more remote areas.*

The fact that the three Peru-forms are commonly imported together does not allow the conclusion that they appear in the same habitats. Usually, all collected fish are brought to the same export station where the animals are gathered until a shipment is worth the trouble. Only because the Peru-Altum is so popular in the hobby, wild caughts of other forms are -accidentally- imported, too.

I hope that in the future enthusiasts will keep and breed more wild forms of the discus and the angelfish than they do at the moment. It would help not only science to (finally) completely understand the systematics of these two genera but also to prevent the original forms from dying out. Because you can only protect what you know. And until today, we know only the domesticated fish - their wild ancestry is still a mystery to us.

Südostasien - Europa
Unterschiede in der Diskus-Zucht

von Manfred Göbel

Diskus-Zuchtanlagen in Asien werden logischerweise in jenen Regionen aufgebaut, welche die besten äußeren Voraussetzungen für eine erfolgreiche Massenvermehrung der Diskusfische versprechen. Wasserqualität, Wasserkosten und Personalkosten sind die ausschlaggebenden Faktoren. Lufttemperatur - und damit auch die Temperatur des Aquarienwassers - liegen in den tropischen Regionen sowieso fast immer im optimalen Bereich.

Technische Probleme können gering gehalten werden da man allgemein mit einer sehr einfachen Technik auskommt. Auf Heizung, künstliche Beleuchtung und Filterung des Aquarienwassers kann fast vollständig verzichtet werden. Ein täglicher Wasserwechsel von bis zu 100 % (und manchmal mehr) ersetzt jede Filterung. Überlegungen zur Wasserchemie können unterbleiben sofern das zur Verfügung stehende Wasser einigermaßen den Bedürfnissen der Fische entspricht (und das ist an den Standorten der Zuchtanlagen fast immer der Fall). Der häufige starke Wasserwechsel sorgt dafür, daß es in den Aquarien zu keiner größeren Verschiebung der Wasserwerte kommen kann. Wozu sich dann Gedanken über Wasserchemie machen?

Die kaufmännische Kalkulation für eine solche Anlage sieht in vielen Regionen Südostasiens zunächst auch sehr positiv aus. Allerdings fehlt dabei oftmals eine Marktanalyse. Das führt dann zu einer völligen Fehleinschätzung des Marktes und der Mitbewerber.

In den letzten Jahren entstand so auf dem Weltmarkt ein riesiges Überangebot an Diskusfischen, das zu einem ruinösen Preiskampf führte. Daß darunter natürlich auch die Qualität der Diskus-Nachzuchten leidet, muß man nicht erst betonen. Aber, „Keine Regel ohne Ausnahme"! Und es gibt sie, auch und gerade in Südostasien, diese „Ausnahme-Zuchtanlagen" für Diskusfische in höchster Qualität. Geführt werden diese Anlagen fast immer von Leuten, die einerseits eine fundierte wissenschaftliche Ausbildung haben, zudem den Markt sehr gut einschätzen können und, was die Hauptsache ausmacht, ein exzellentes Gefühl für die züchterischen Möglichkeiten bei Diskusfischen besitzen.

Im Gegensatz zu den allermeisten asiatischen Diskus-Großzüchtereien, die von Anfang an auf rein kommerzieller Basis entstanden sind, hat die Diskus-Zucht in Europa und hier vor allem in Deutschland fast immer einen ideellen Hintergrund.

Generell kann man feststellen, daß Menschen, die in einem kälteren Klima leben, ein anderes Verhältnis zu Tieren entwickeln, als Menschen in tropischen Regionen. Trübe Herbsttage und kalte Winter führen dazu,

Southeast Asia and Europe - Differences in Commercial Discus Breeding

By Manfred Göbel

In Asia, the large commercial discus breeding farms are, of course, set up in regions where all necessary requirements for successful mass reproduction of ornamental fish are present.

Water quality and the costs for water and workers are the most important factors. In the tropical regions of Southeast Asia, the air temperature, and, consequently, the temperature of the aquarium water are within the optimum range anyway.

Due to this favourable climate, the technical problems of supporting such an installation play only a minor role; usually, very simple technical devices are used. Heating systems, illumination and filtration can be almost completely neglected. A daily water change of up to 100% of the water (and sometimes even several times a day!) makes any kind of filtration unnecessary. Water chemistry is no issue either, as long as the parameters of the 'tap' water are roughly the ones required by the fish (and this is mostly the case in these countries). The frequent large water changes make sure that the water chemistry stays constant - so why bother about it?

The economic calculation for setting up such a farm in Southeast Asia looks, at first, just as favourable. But very often, no market analysis is carried through beforehand, so that the market and the competition are totally misjudged. In the last years, an enormous surplus of discus flooded on the world trade market which led to a ruinous drop in prices; and, of course, a decline in quality in many commercial breeds. Still, there is no rule without exception, as there are indeed exceptional, high-quality farms breeding discus in Southeast Asia. These high standard farms are usually managed by people with a fundamental education as well as an insight into the laws of the market and, most importantly, the necessary "feeling" for the possibilities of breeding high-quality discus strains.

In contrast to the majority of Asian discus farms that are based on a solely commercial basis, in Europe and Germany, the breeding of discus has mostly an idealistic background. Generally, one can say that people living in areas with a colder climate have a different relationship with animals than inhabitants of tropical regions.

daß die Menschen viel Zeit im Haus und damit ohne direkten Kontakt zur freien Natur verbringen. Was liegt da näher, als der Versuch, die Natur ins Haus zu holen. Da dies aber viel Sorgfalt und einen relativ hohen technischen Aufwand bedeutet, entwickeln die Menschen dabei häufig ein recht inniges Verhältnis zu den gepflegten Pflanzen und Tieren.

Diese besonders enge Bindung zwischen Pfleger und Pflegling bildet auch den ideellen Hintergrund für die Pflege von Diskusfischen und deren Zucht in Nordeuropa. Fast immer ist es zunächst der interessierte Hobby-Aquarianer, der sich einige Fische besorgt um sie zu pflegen und zu beobachten. Oft sind die äußeren Voraussetzungen alles andere als ideal. Das zur Verfügung stehend Wasser ist häufig für die Pflegen der Diskusfische nur bedingt geeignet. Also muß es aufbereitet werden. Außerdem sind die Wasser- und Abwasserkosten im Vergleich zu fast allen asiatischen Ländern sehr hoch. Das zwingt den Aquarianer in Europa, sich von Anfang an intensiv mit Wasserchemie, Aquarientechnik und Filterung zu befassen.

Der technische Aufwand wächst meistens beträchtlich wenn man seine Pflegling zur Nachzucht bringen will. Die Voraussetzungen für eine entstehende Diskuszucht in Nordeuropa sind also beileibe nicht Gewinnerwartung und ideale Voraussetzungen, sondern Neugierde und Interesse des Aquarianers an seinen Fischen.

Wenn bei einem Hobby-Aquarianer zum ersten Mal ein Diskus-Paar Laichvorbereitungen trifft, so ist das für diesen Aquarianer auch heute noch ein besonders Ereignis, das eine ganze Reihe von Aktivitäten auslösen kann. Ein Zuchtbecken wird besorgt und installiert, spezielles „Zuchtwasser" wird aufbereitet und das Diskus-Paar wird in diese „Idylle" umgesetzt. Nicht selten kann der Aquarianer einige Zeit später zum ersten Mal die Umsicht und Sorgfalt bewundern, mit der die Diskus-Eltern ihre erste Brut pflegen und aufziehen.

Überwiegen bei diesem Aquarianer kommerzielle Interessen, so wird er sich möglicherweise in naher Zukunft in den Kreis der Diskus-Vermehrer einreihen. Überwiegen jedoch Staunen, Bewunderung und Neugierde, so wird er sich bestimmt darauf besinnen, was der Begriff „Zucht" eigentlich bedeutet. Im Laufe der Zeit wird er bei seiner Arbeit mit den Diskusfischen bestimmte Ziele definieren und bestrebt sein, diese Zuchtziele auch zu erreichen. Dazu werden mehr Aquarien benötigt. Die technischen Voraussetzungen werden verbessert und Abläufe werden - soweit möglich - automatisiert, um die vergrößerte Anlage mit vertretbarem Zeitaufwand zu betreiben. Aus wenigen Aquarien im Anfang entsteht so allmählich eine kleine Diskus-Zuchtanlage. Natürlich betrachtet der Aquarianer diese Anlage noch immer als sein Hobby, obwohl er täglich einen Großteil seiner Zeit dafür aufwendet. Familienmitglieder mögen ab und zu helfen, die anfallenden Arbeiten zu erledigen. Angestell-

Rainy autumns and cold winters make people stay inside their houses so that there is no intense contact with nature for a rather long period of the year. So what seems more sensible that to get nature - animals and plants - inside the house? The arising need to maintain the sensitive co-inhabitants and the resulting effort and technical input prompts keepers of animals and plants to devote a lot of time to their "pets". Consequently, many hobbyists have a profound relationship with the "objects" they take care of. This intimate relationship of keeper and "ward" is the non-material background of keeping and breeding discus in Northern Europe and Germany. It is basically the interest hobbyist who buys discus to keep and observe them. Very often the conditions for keeping fish are far from ideal. The available water is, mostly, only partly suited for breeding and keeping discus. As a consequence, the water has to be conditioned. Also, the costs for water and sewage are much higher than in Asia. This forces European aquarists right from the beginning to intensely deal with water chemistry, aquarium equipment and filtration systems.

The technical effort even has to be increased when hobbyists want to spawn their discus. Thus, the reasons for a future discus breeding installation are not profit or the easiness with which such an installation is et up, but sheer enthusiasm. The first time, a hobbyist prepares for breeding his discus is always a special moment in an "aquatic career". The aquarist becomes very busy: A breeding tank is bought and set up, special "breeding water" is prepared and the parent fish are introduced to their new home. And - very often even the very first attempts are indeed successful and the fortunate aquarist can admire his pair of discus caring tenderly for their fry.

When a n aquarist is predominantly interested in profit, he might become one of the many discus "producers". But when enthusiasm and admiration are the main points of interest, hobby aquarists will certainly recognise what the word "breeding" really means. In the course of time, the enthusiast will define certain goals he wants to reach as a breeder. For this, he needs more than one or two aquaria. The equipment is updated and the daily maintenance routine is automated to carry it through with as little time effort as possible. This way, slowly a small breeding installation develops from the first aquarium. Of course, the hobbyist still regards this installation as pure "hobby" although he might spend most of his time breeding and tending his fish. Family members may be integrated to help with the daily maintenance routine, but one will never find any

te Arbeitskräfte wird man in diesen Zuchtanlagen nicht finden.

So entscheidet die Leistungsfähigkeit des einzelnen Aquarianers und die täglich zur Verfügung stehende Zeit letztendlich über die Größe einer solchen Anlage. 50 bis 60 Aquarien sind oft das Maximum. Häufig sind die Zuchtanlagen wesentlich kleiner.

Die relativ geringe Anzahl der zur Verfügung stehenden Aquarien führt fast immer dazu, daß man sich bei der züchterischen Arbeit auf wenige Farbschläge beschränkt und bei der Auswahl der Zuchttiere sehr sorgfältig vorgeht. Natürlich sind auch in Europa im Laufe der Zeit einige große Diskus-Zuchtbetriebe entstanden. Doch der Großteil der gezüchteten Diskusfische kommt nach wie vor aus den beschriebenen kleinen Zuchtanlagen. Diese Tiere sind oftmals von hervorragender Qualität und finden fast immer ihre Abnehmer im Markt.

Da für die Züchter die Tiere im Mittelpunkt stehen, haben Gesundheit und Wohlbefinden der Fische oberste Priorität. Die eingesetzte Technik, das verwendete Wasser und das Futter muß den Bedürfnissen der Tiere entsprechen. Nur optimale Pflege bietet eine gute Möglichkeit, Krankheiten und Parasitenbefall weitestgehend zu vermeiden.

Sollten sich dennoch Parasiten verstärkt vermehren oder sollte gar eine ernsthafte Krankheit auftreten, so wird man das in der Regel frühzeitig erkennen, wenn man seine Diskusfische mit der gebührenden Sorgfalt beobachtet.

Eine genaue Diagnose der Erkrankung muß auf jeden Fall einer Behandlung mit Medikamenten oder Chemikalien vorausgehen, da diese Stoffe die Fische und das Wasser zusätzlich stark belasten. Und mit erkrankten oder durch Chemikalien geschädigten Diskusfischen läßt sich nun mal keine gesunde Zucht aufbauen.

Verantwortungsbewußtsein, Einfühlungsvermögen, die Liebe zum Tier und viel Geduld sind die Grundpfeiler jeder guten Diskuszucht. In vielen Ländern der Erde werden in kleinen und großen Anlagen Diskusfische gezüchtet, die uns die ganze Harmonie und Schönheit dieser Tiere zeigen und somit dazu beitragen, daß uns auch in Zukunft die „Faszination Diskus" erhalten bleibt.

employees in such an "amateur" breeding farm. Owing to these circumstances, the hobbyist's abilities and energies are the decisive factors for the size of the breeding installation. 50 to 60 tanks are the maximum number you will find here; most of the time, these hobby breeding installations are much smaller. The reduced space and number of aquaria leads almost inevitably to a very careful selection of colour strains and breeding specimens. Of course, in Europe, too, large commercial breeding farms have established. But the majority of discus bred in Europe come indeed from the small installations described above. These discus are always high-quality fish that are easily sold.

To hobbyists, the health and well-being of their fish is of top priority. The equipment, the water and the foods have to be of the highest possible standard. Only optimal care guarantees an almost disease and parasite free aquarium. Parasites infesting the tanks or even serious disease break-outs despite careful maintenance will certainly be recognised at a every early stage, if the aquarist observes his fish attentively. An exact diagnosis of given disease is necessary before any medications or treatments are applied, because any added chemicals strain the fish and the plants living in the aquarium. And breeding with stressed or even ill animals is simply not possible.

Sense of responsibility, sensibility, love of animals and a lot of patience are the cornerstones of any high-quality discus breeding. In many countries of the world, discus are bred in small and large breeding farms; the beauty and charm of the majestic discus fish are preserved this way so that in the future, these fascinating fish will stay with us and brighten up our tanks and lives.

Symbole

Um möglichst alle Fische im Bild zeigen zu können und um dem weltweiten Vertrieb unserer Bildbände Rechnung zu tragen, haben wir bewußt auf ausführliche Texte verzichtet und ersetzen diese durch internationale Symbole, mit deren Hilfe jeder leicht die wichtigsten Eigenschaften der Fische und deren Pflege erkennen kann.

Ursprung:

ersehen Sie ganz leicht an dem Buchstaben vor der Code-Nummer

- **A** = Afrika
- **E** = Europa + Nordamerika
- **S** = Südamerika
- **X** = Asien + Australien

Alter:

die letzte Zahl der Code-Nummer steht immer für das Alter des fotografierten Fisches:

- **1** = klein (Jugendfärbung)
- **2** = mittelgroß (Jungfisch / juvenil / Verkaufsgröße)
- **3** = groß (halbwüchsig / gute Verkaufsgröße)
- **4** = XL (ausgewachsen / adult)
- **5** = XXL (Zucht-Tier)
- **6** = show (Schau-Tier)

Herkunft:

- **W** = Wildform
- **B** = Nachzucht
- **Z** = Zuchtform
- **X** = Kreuzungs-Form

Größe:

..cm = ungefähre Größe, die dieser Fisch ausgewachsen (adult) erreichen kann.

Geschlecht:

♂ männlich   ♀ weiblich   ♂♀ Paar

Temperatur:

- ◁ 18-22°C (64 - 72°F) (Zimmertemperatur)
- ▷ 22-25°C (72 - 77°F) (tropische Fische)
- △ 24-29°C (75 - 85°F) (Diskus etc.)
- ▽ 10-22°C (50 - 72°F) (kalt)

pH-Wert:

- ₧ pH 6,5 - 7,2 keine besonderen Ansprüche (neutral)
- ⇩P pH 5,8 - 6,5 liebt weiches u. leicht saures Wasser
- ⇧P pH 7,5 - 8,5 liebt hartes u. alkalisches Wasser

Beleuchtung:

- ○ hell, viel Licht / Sonne
- ◐ nicht zu hell
- ● fast dunkel

Futter:

- ☺ Allesfresser, Trockenfutter, keine besonderen Ansprüche
- ☻ Lebendfutter, Gefrierfutter
- ☹ Fischräuber, Futterfische füttern
- ☻ Pflanzenfresser, Pflanzenkost zufüttern

Schwimmverhalten:

- ⊞ keine besonderen Eigenschaften
- ⊤ im oberen Bereich / Oberflächenfisch
- ⊥ im unteren Bereich / Bodenfisch

Aquarium-Einrichtung:

- 🗔 nur Bodengrund und Steine etc.
- 🗔 Steine / Wurzeln / Höhlen
- 🗔 Pflanzen-Aquarium mit Dekoration

Verhalten/Vermehrung:

- ♥ Paarweise oder im Trio halten
- 🐟 Schwarmfisch, nicht unter 10 Exemplaren halten
- 🐟 Eierleger
- 🐟 Lebendgebärer
- 🐟 Maulbrüter
- 🐟 Höhlenbrüter
- 🐟 Schaumnestbauer
- ◯ Algenvertilger / Scheibenputzer (Wurzeln+Spinat)
- ◇ leichte Pflege (für entsprechende Gesellschaftsbecken)
- ⚠ schwierig zu halten, vorher Fachliteratur beachten
- 🛑 Vorsicht, extrem schwierig, nur für erfahrene Spezialisten
- 0 die Eier benötigen eine spezielle Behandlung
- § geschützte Art, (WA), "CITES" Sondergenehmigung nötig

Mindestgröße des Aquariums / Inhalt:

Symbol	Größe	Maße	Inhalt
ss	sehr klein	20 - 40 cm	5 - 20 l
s	klein	40 - 80 cm	40 - 80 l
m	mittel	60 - 100 cm	80 - 200 l
L	groß	100 - 200 cm	200 - 400 l
xL	sehr groß	200 - 400 cm	400 - 3000 l
xxL	extrem groß	über 400 cm	über 3000 l
			(Schauaquarien)

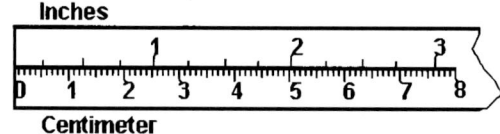
Inches / Centimeter

Gruppe 1 / Group 1: Discus Heckel

S89701-4 *Symphysodon discus* Heckel, 1840
DISCUS HECKEL
Brazil, wild

Photo: H.J. Mayland

S89702-4 *Symphysodon discus* Heckel, 1840
DISCUS HECKEL
Brazil, wild

Photo: H.J. Mayland

S89703-4 *Symphysodon discus* Heckel, 1840
DISCUS HECKEL
Brazil, wild

Photo: H.J. Mayland

S89704-4 *Symphysodon discus* Heckel, 1840
DISCUS HECKEL
Brazil, wild

Photo: H.J. Mayland

S89705-4 *Symphysodon discus* Heckel, 1840
DISCUS HECKEL
Brazil, wild

Photo: Schlingmann

S89706-4 *Symphysodon discus* Heckel, 1840
DISCUS HECKEL
Brazil, wild

Photo: Archiv A.C.S.

Gruppe 1 / Group 1: Discus Heckel

S89707-4 *Symphysodon discus* HECKEL, 1840
DISCUS HECKEL
Brazil, wild

Photo: M. Göbel

S89708-4 *Symphysodon discus* HECKEL, 1840
DISCUS HECKEL
Brazil, wild

Photo: M. Göbel

S89709-4 *Symphysodon discus* HECKEL, 1840
DISCUS HECKEL
Brazil, wild

Photo: M. Göbel

S89710-4 *Symphysodon discus* HECKEL, 1840
DISCUS HECKEL
Brazil, wild

Photo: H.J. Mayland

S89711-4 *Symphysodon discus* HECKEL, 1840
DISCUS HECKEL
Brazil, wild

Photo: F. Bodenmüller

S89712-4 *Symphysodon discus* HECKEL, 1840
DISCUS HECKEL
Brazil, wild

Photo: F. Bodenmüller

Gruppe 1 / Group 1: Discus Heckel

S89801-4 *Symphysodon discus* Heckel, 1840
DISCUS HECKEL-Blaukopf/Bluehead
Brazil, wild

Photo: Nakano/Archiv A.C.S.

S89802-4 *Symphysodon discus* Heckel, 1840
DISCUS HECKEL-Blaukopf/Bluehead
Brazil, wild

Photo: H.J. Mayland

S89803-4 *Symphysodon discus* Heckel, 1840
DISCUS HECKEL-Blaukopf/Bluehead
Brazil, wild

Photo: M.Tomizana/A.C.S.

S89804-4 *Symphysodon discus* Heckel, 1840
DISCUS HECKEL-Blaukopf/Bluehead
Brazil, wild

Photo: Nakano/Archiv A.C.S.

S89805-4 *Symphysodon discus* Heckel, 1840
DISCUS HECKEL-Blaukopf/Bluehead
Brazil, wild

Photo: M. Göbel

S89806-4 *Symphysodon discus* Heckel, 1840
DISCUS HECKEL-Blaukopf/Bluehead
Brazil, wild

Photo: M. Göbel

South American **Cichlids IV**

S89810-5 *Symphysodon discus* Heckel, 1840
DISCUS HECKEL Brazil, wild   Männchen / Male   Normalfärbung / neutral colour
Photo: J. Schütz

S89817-5 *Symphysodon discus* Heckel, 1840   (das gleiche Männchen wie auf der nächsten Seite / the same male as on the next page)
DISCUS HECKEL Brazil, wild   Männchen / Male   Normalfärbung / neutral colour
Photo: J. Schütz

S89817-5 *Symphysodon discus* Heckel, 1840
DISCUS HECKEL Brazil, wild   Paar in Brutstimmung / pair in breeding mood   Photo: J. Schütz

S89817-5 *Symphysodon discus* Heckel, 1840
DISCUS HECKEL Brazil, wild   Paar in Brutstimmung / pair in breeding mood   Photo: J. Schütz

Gruppe 1 / Group 1: Discus Heckel

S89810 *Symphysodon discus* Heckel, 1840
DISCUS HECKEL
beim Ablaichen / spawning
Photo: J. Schütz

S89810 *Symphysodon discus* Heckel, 1840
DISCUS HECKEL
Weibchen mit Jungfischen / Female with babies
Photo: J. Schütz

S89810 *Symphysodon discus* Heckel, 1840
DISCUS HECKEL
Weibchen mit Jungfischen / Female with babies
Photo: J. Schütz

S89810 *Symphysodon discus* Heckel, 1840
DISCUS HECKEL
Schwarm Jungfische / school of young fish
Photo: J. Schütz

S89810 *Symphysodon discus* Heckel, 1840
DISCUS HECKEL
Jungfisch, 5-6 Wochen alt / young fish, 5-6 weeks old
Photo: J. Schütz

S89810 *Symphysodon discus* Heckel, 1840
DISCUS HECKEL
Jungfisch, 10 Wochen alt / young fish, 10 weeks old
Photo: J. Schütz

S89811-4 *Symphysodon discus* Heckel, 1840
F$_1$-Generation, ca. 2 Jahre alt / about 2 years old; alle Tiere auf dieser Seite sind Geschwister / all specimen on this page are brothers and sisters
Photo: J. Schütz

S89812-4 *Symphysodon discus* Heckel, 1840
F$_1$-Generation, ca. 2 Jahre alt / about 2 years old; alle Tiere auf dieser Seite sind Geschwister / all specimen on this page are brothers and sisters
Photo: J. Schütz

S89813-4 *Symphysodon discus* Heckel, 1840
F$_1$-Generation, ca. 2 Jahre alt / about 2 years old; alle Tiere auf dieser Seite sind Geschwister / all specimen on this page are brothers and sisters
Photo: J. Schütz

S89814-4 *Symphysodon discus* Heckel, 1840
F$_1$-Generation, ca. 2 Jahre alt / about 2 years old; alle Tiere auf dieser Seite sind Geschwister / all specimen on this page are brothers and sisters
Photo: J. Schütz

S89815-4 *Symphysodon discus* Heckel, 1840
F$_1$-Generation, ca. 2 Jahre alt / about 2 years old; alle Tiere auf dieser Seite sind Geschwister / all specimen on this page are brothers and sisters
Photo: J. Schütz

S89816-4 *Symphysodon discus* Heckel, 1840
F$_1$-Generation, ca. 2 Jahre alt / about 2 years old; alle Tiere auf dieser Seite sind Geschwister / all specimen on this page are brothers and sisters
Photo: J. Schütz

S89820-4 *Symphysodon* Hybride
Vater / Father: Heckel wild x Rottürkis / Red Turqoise, Mutter / Mother: Heckel wild x (Heckel wild x Rottürkis / Red Turqoise)
Photo: M. Göbel

S89825-4 *Symphysodon* Hybride
Kreuzung von Heckel-Männchen und rot-türkis Weibchen / cross-breed Heckel (male) x red-turquoise (female)

Photo: J. Schütz

Gruppe 2 / Group 2: Discus Brown

S89301-4 *Symphysodon aequifasciatus* Pellegrin, 1904
DISCUS BROWN, „Rio Madeira"
Brazil, wild

Photo: H.J. Mayland

S89302-4 *Symphysodon aequifasciatus* Pellegrin, 1904
DISCUS BROWN, „Rio Madeira"
Brazil, wild

Photo: U. Werner

S89303-4 *Symphysodon aequifasciatus* Pellegrin, 1904
DISCUS BROWN, „Rio Madeira"
Brazil, wild

Photo: U. Werner

S89304-4 *Symphysodon aequifasciatus* Pellegrin, 1904
DISCUS BROWN, „Rio Madeira"
Brazil, wild

Photo: U. Werner

S89305-4 *Symphysodon aequifasciatus* Pellegrin, 1904
DISCUS BROWN, „Rio Madeira"
Brazil, wild

Photo: M. Göbel

S89306-4 *Symphysodon aequifasciatus* Pellegrin, 1904
DISCUS BROWN, „Rio Madeira"
Brazil, wild

Photo: M. Göbel

S89307-4 *Symphysodon aequifasciatus* Pellegrin, 1904
DISCUS BROWN, „Rio Madeira"
Brazil, wild

Photo: U. Werner

S89308-4 *Symphysodon aequifasciatus* Pellegrin, 1904
DISCUS BROWN, „Rio Madeira"
Brazil, wild

Photo: U. Werner

S89309-4 *Symphysodon aequifasciatus* Pellegrin, 1904
DISCUS BROWN, „Rio Madeira"
Brazil, wild

Photo: W. Mikschofsky

S89310-4 *Symphysodon aequifasciatus* Pellegrin, 1904
DISCUS MADEIRAL
Brazil, wild

Photo: M. Göbel

S89311-4 *Symphysodon aequifasciatus* Pellegrin, 1904
DISCUS BROWN, „Rio Madeira"
Brazil, wild

Photo: M. Göbel

S89312-4 *Symphysodon aequifasciatus* Pellegrin, 1904
DISCUS BROWN, „Rio Madeira"
Brazil, wild

Photo: M. Göbel

Gruppe 2 / Group 2: Discus Brown

S89313-4 *Symphysodon aequifasciatus* Pellegrin, 1904
DISCUS BROWN, „Rio Madeira"
Brazil, wild

Photo: H.J. Mayland

S89314-4 *Symphysodon aequifasciatus* Pellegrin, 1904
DISCUS BROWN, „Rio Madeira"
Brazil, wild

Photo: H.J. Mayland

S89315-4 *Symphysodon aequifasciatus* Pellegrin, 1904
DISCUS BROWN, „Rio Madeira"
Brazil, wild

Photo: H.J. Mayland

S89316-4 *Symphysodon aequifasciatus* Pellegrin, 1904
DISCUS BROWN, „Rio Madeira"
Brazil, wild

Photo: F. Bodenmüller

S89317-4 *Symphysodon aequifasciatus* Pellegrin, 1904
DISCUS BROWN, „Rio Madeira"
Brazil, wild

Photo: H.J. Mayland

S89318-4 *Symphysodon aequifasciatus* Pellegrin, 1904
DISCUS BROWN, „Rio Madeira"
Brazil, wild

Photo: Nakano/Archiv A.C.S.

S89319-4 *Symphysodon aequifasciatus* PELLEGRIN, 1904
DISCUS BROWN, „Rio Madeira"
Brazil, wild

Photo: H.J. Mayland

S89320-4 *Symphysodon aequifasciatus* PELLEGRIN, 1904
DISCUS BROWN, „Rio Madeira"
Brazil, wild

Photo: M. Göbel

S89321-4 *Symphysodon aequifasciatus* PELLEGRIN, 1904
DISCUS BROWN, „Rio Madeira"
Brazil, wild

Photo: H.J. Mayland

S89322-4 *Symphysodon aequifasciatus* PELLEGRIN, 1904
DISCUS BROWN, „Rio Madeira"
Brazil, wild

Photo: M. Göbel

S89323-4 *Symphysodon aequifasciatus* PELLEGRIN, 1904
DISCUS BROWN, „Rio Madeira"
Brazil, wild

Photo: H.J. Mayland

S89324-4 *Symphysodon aequifasciatus* PELLEGRIN, 1904
DISCUS BROWN, „Rio Madeira"
Brazil, wild

Photo: F. Bodenmüller

Gruppe 2 / Group 2: Discus Brown

S89325-4 *Symphysodon aequifasciatus* Pellegrin, 1904
DISCUS BROWN, „Rio Madeira"
Brazil, wild

Photo: F. Bodenmüller

S89326-4 *Symphysodon aequifasciatus* Pellegrin, 1904
DISCUS BROWN, „Rio Madeira"
Brazil, wild

Photo: W. Mikschofsky

S89327-4 *Symphysodon aequifasciatus* Pellegrin, 1904
DISCUS BROWN, „Rio Madeira"
Brazil, wild

Photo: Nakano / Archiv A.C.S.

S89328-4 *Symphysodon aequifasciatus* Pellegrin, 1904
DISCUS BROWN, „Rio Madeira"
Brazil, wild

Photo: W. Mikschofsky

S89329-4 *Symphysodon aequifasciatus* Pellegrin, 1904
DISCUS BROWN, „Rio Madeira" "GELB/YELLOW"
Brazil, wild

Photo: M. Göbel

S89329-4 *Symphysodon aequifasciatus* Pellegrin, 1904
DISCUS BROWN, „Rio Madeira" "GELB/YELLOW"
Brazil, wild

Photo: M. Göbel

Gewinner des ersten internationalen Diskuschampionats 1996 in Duisburg

Best of show at the first International Discus Championship 1996 in Duisburg/Germany

S89300-4 *Symphysodon aequifasciatus* PELLEGRIN, 1904
DISCUS BROWN, „Rio Madeira"
Brazil, wild

Photo: Fa. Zoo Zajac

Gruppe 2 / Group 2: Discus Brown

S89331-4 *Symphysodon aequifasciatus* Pellegrin, 1904
DISCUS BROWN, „Rio Madeira"
Brazil, wild

Photo: M. Göbel

S89332-4 *Symphysodon aequifasciatus* Pellegrin, 1904
DISCUS BROWN, „Rio Madeira"
Brazil, wild

Photo: F. Bodenmüller

S89333-4 *Symphysodon aequifasciatus* Pellegrin, 1904
DISCUS BROWN, „Rio Madeira"
Brazil, wild

Photo: F. Bodenmüller

S89334-4 *Symphysodon aequifasciatus* Pellegrin, 1904
DISCUS BROWN, „Rio Madeira"
Brazil, wild

Photo: Nakano / Archiv A.C.S.

S89335-4 *Symphysodon aequifasciatus* Pellegrin, 1904
DISCUS BROWN, „Rio Madeira"
Brazil, wild

Photo: F. Bodenmüller

S89336-4 *Symphysodon aequifasciatus* Pellegrin, 1904
DISCUS BROWN, „Rio Madeira"
Brazil, wild

Photo: F. Bodenmüller

S89337-4 *Symphysodon aequifasciatus* Pellegrin, 1904
DISCUS BROWN, „Rio Madeira"
Brazil, wild

Photo: F. Bodenmüller

S89338-4 *Symphysodon aequifasciatus* Pellegrin, 1904
DISCUS BROWN, „Rio Madeira"
Brazil, wild

Photo: F. Bodenmüller

S89339-4 *Symphysodon aequifasciatus* Pellegrin, 1904
DISCUS BROWN, „Rio Madeira"
Brazil, wild

Photo: M. Göbel

S89340-4 *Symphysodon aequifasciatus* Pellegrin, 1904
DISCUS BROWN, „Rio Madeira"
Brazil, wild

Photo: F. Bodenmüller

S89341-4 *Symphysodon aequifasciatus* Pellegrin, 1904
DISCUS BROWN, „Rio Madeira"
Brazil, wild

Photo: Colle/Archiv A.C.S.

S89342-4 *Symphysodon aequifasciatus* Pellegrin, 1904
DISCUS BROWN, „Rio Madeira"
Brazil, wild

Photo: F. Bodenmüller

Gruppe 2 / Group 2: Discus Brown

S89343-4 *Symphysodon aequifasciatus* Pellegrin, 1904
DISCUS BROWN, „Rio Madeira"
Brazil, wild

Photo: F. Bodenmüller

S89344-4 *Symphysodon aequifasciatus* Pellegrin, 1904
DISCUS BROWN, „Rio Madeira"
Brazil, wild

Photo: F. Bodenmüller

S89345-4 *Symphysodon aequifasciatus* Pellegrin, 1904
DISCUS BROWN, „Rio Madeira"
Brazil, wild

Photo: F. Bodenmüller

S89346-4 *Symphysodon aequifasciatus* Pellegrin, 1904
DISCUS BROWN, „Rio Madeira"
Brazil, wild

Photo: F. Bodenmüller

S89347-4 *Symphysodon aequifasciatus* Pellegrin, 1904
DISCUS BROWN, „Rio Madeira"
Brazil, wild

Photo: F. Bodenmüller

S89348-4 *Symphysodon aequifasciatus* Pellegrin, 1904
DISCUS BROWN, „Rio Madeira"
Brazil, wild

Photo: F. Bodenmüller

S89349-4 *Symphysodon aequifasciatus* Pellegrin, 1904
DISCUS BROWN, „Rio Madeira"
Brazil, wild

Photo: M. Göbel

S89350-4 *Symphysodon aequifasciatus* Pellegrin, 1904
DISCUS BROWN, „Rio Madeira"
Brazil, wild

Photo: M. Göbel

S89351-4 *Symphysodon aequifasciatus* Pellegrin, 1904
DISCUS BROWN, „Rio Madeira"
Brazil, wild

Photo: F. Bodenmüller

S89352-4 *Symphysodon aequifasciatus* Pellegrin, 1904
DISCUS BROWN, „Rio Madeira"
Brazil, wild

Photo: M. Göbel

S89353-4 *Symphysodon aequifasciatus* Pellegrin, 1904
DISCUS BROWN, „Rio Madeira"
Brazil, wild

Photo: M. Göbel

S89354-4 *Symphysodon aequifasciatus* Pellegrin, 1904
DISCUS BROWN, „Rio Madeira"
Brazil, wild

Photo: M. Göbel

Gruppe 2 / Group 2: Discus Brown

S89355-4 *Symphysodon aequifasciatus* Pellegrin, 1904
DISCUS BROWN, „Rio Madeira"
Brazil, wild

Photo: F. Bodenmüller

S89356-4 *Symphysodon aequifasciatus* Pellegrin, 1904
DISCUS BROWN, „Rio Madeira"
Brazil, wild

Photo: F. Bodenmüller

S89357-4 *Symphysodon aequifasciatus* Pellegrin, 1904
DISCUS BROWN, „Rio Madeira"
Brazil, wild

Photo: U. Werner

S89358-4 *Symphysodon aequifasciatus* Pellegrin, 1904
DISCUS BROWN, „Rio Madeira"
Brazil, wild

Photo: Nakano / Archiv A.C.S.

S89359-4 *Symphysodon aequifasciatus* Pellegrin, 1904
DISCUS BROWN, „Rio Madeira"
Brazil, wild

Photo: U. Werner

S89360-4 *Symphysodon aequifasciatus* Pellegrin, 1904
DISCUS BROWN, „Rio Madeira"
Brazil, wild

Photo: U. Werner

S89361-4 *Symphysodon aequifasciatus* Pellegrin, 1904
DISCUS BROWN, „Rio Madeira"
Brazil, wild

Photo: M. Göbel

S89362-4 *Symphysodon aequifasciatus* Pellegrin, 1904
DISCUS BROWN, „Rio Madeira"
Brazil, wild

Photo: M. Göbel

S89363-4 *Symphysodon aequifasciatus* Pellegrin, 1904
DISCUS BROWN, „Rio Madeira"
Brazil, wild

Photo: M. Göbel

S89364-4 *Symphysodon aequifasciatus* Pellegrin, 1904
DISCUS BROWN, „Rio Madeira"
Brazil, wild

Photo: M. Göbel

S89365-4 *Symphysodon aequifasciatus* Pellegrin, 1904
DISCUS BROWN, „Rio Madeira"
Brazil, wild

Photo: F. Bodenmüller

S89366-4 *Symphysodon aequifasciatus* Pellegrin, 1904
DISCUS BROWN, „Rio Madeira"
Brazil, wild

Photo: F. Bodenmüller

Gruppe 2 / Group 2: Discus Brown

S89401-4 *Symphysodon aequifasciatus* Pellegrin, 1904
DISCUS BROWN „ALENQUER"
Brazil, wild

Photo: W. Mikschofsky

S89402-4 *Symphysodon aequifasciatus* Pellegrin, 1904
DISCUS BROWN „ALENQUER"
Brazil, wild

Photo: W. Mikschofsky

S89403-4 *Symphysodon aequifasciatus* Pellegrin, 1904
DISCUS BROWN „ALENQUER"
Brazil, wild

Photo: M. Göbel

S89404-4 *Symphysodon aequifasciatus* Pellegrin, 1904
DISCUS BROWN „ALENQUER", PAIR
Brazil, wild

Photo: W. Mikschofsky

S89405-4 *Symphysodon aequifasciatus* Pellegrin, 1904
DISCUS BROWN „ALENQUER"
Brazil, wild

Photo: W. Mikschofsky

S89406-4 *Symphysodon aequifasciatus* Pellegrin, 1904
DISCUS BROWN „ALENQUER"
Brazil, wild

Photo: W. Konrad

South American **Cichlids** IV   Verlag A.C.S. GmbH

S89365-4 *Symphysodon aequifasciatus* Pellegrin, 1904
DISCUS BROWN, „Rio Madeira"
Brazil, wild

S89406-4 *Symphysodon aequifasciatus* Pellegrin, 1904
DISCUS BROWN „ALENQUER"
Brazil, wild

S89407-4 *Symphysodon aequifasciatus* Pellegrin, 1904
DISCUS BROWN „ALENQUER"
Brazil, wild

Photo: W. Mikschofsky

S89408-4 *Symphysodon aequifasciatus* Pellegrin, 1904
DISCUS BROWN „ALENQUER"
Brazil, wild

Photo: H.J. Mayland

S89409-4 *Symphysodon aequifasciatus* Pellegrin, 1904
DISCUS BROWN „ALENQUER"
Brazil, wild

Photo: W. Mikschofsky

S89410-4 *Symphysodon aequifasciatus* Pellegrin, 1904
DISCUS BROWN „ALENQUER"
Brazil, wild

Photo: W. Mikschofsky

S89411-4 *Symphysodon aequifasciatus* Pellegrin, 1904
DISCUS BROWN „ALENQUER"
Brazil, wild

Photo: H.J. Mayland

S89412-4 *Symphysodon aequifasciatus* Pellegrin, 1904
DISCUS BROWN „ALENQUER"
Brazil, wild

Photo: M. Göbel

Gruppe 2 / Group 2: Discus Brown

S89413-4 *Symphysodon aequifasciatus* Pellegrin, 1904
DISCUS BROWN „ALENQUER"
Brazil, wild

Photo: W. Mikschofsky

S89414-4 *Symphysodon aequifasciatus* Pellegrin, 1904
DISCUS BROWN „ALENQUER"
Brazil, wild

Photo: W. Mikschofsky

S89415-4 *Symphysodon aequifasciatus* Pellegrin, 1904
DISCUS BROWN „ALENQUER"
Brazil, wild

Photo: F. Bodenmüller

S89416-4 *Symphysodon aequifasciatus* Pellegrin, 1904
DISCUS BROWN „ALENQUER"
Brazil, wild

Photo: H.J. Mayland

S89417-4 *Symphysodon aequifasciatus* Pellegrin, 1904
DISCUS BROWN „ALENQUER"
Brazil, wild

Photo: H.J. Mayland

S89418-4 *Symphysodon aequifasciatus* Pellegrin, 1904
DISCUS BROWN „ALENQUER"
Brazil, wild

Photo: W. Mikschofsky

S89412-4 *Symphysodon aequifasciatus* Pellegrin, 1904
DISCUS BROWN „ALENQUER"
Brazil, wild

Gruppe 2 / Group 2: Discus Brown

S89419-4 *Symphysodon aequifasciatus* PELLEGRIN, 1904
DISCUS BROWN „ALENQUER", PAIR
Brazil, wild

Photo: H.J. Mayland

S89420-4 *Symphysodon aequifasciatus* PELLEGRIN, 1904
DISCUS BROWN „ALENQUER", PAIR
Brazil, wild

Photo: H.J. Mayland

S89421-4 *Symphysodon aequifasciatus* PELLEGRIN, 1904
DISCUS BROWN „ALENQUER", PAIR
Brazil, wild

Photo: H.J. Mayland

S89422-4 *Symphysodon aequifasciatus* PELLEGRIN, 1904
DISCUS BROWN „ALENQUER"
Brazil, wild

Photo: H.J. Mayland

S89423-4 *Symphysodon aequifasciatus* PELLEGRIN, 1904
DISCUS BROWN „ALENQUER"
Brazil, wild

Photo: H.J. Mayland

S89424-4 *Symphysodon aequifasciatus* PELLEGRIN, 1904
DISCUS BROWN „ALENQUER"
Brazil, wild

Photo: H.J. Mayland

South American Cichlids IV © Verlag A.C.S. GmbH

S89425-4 *Symphysodon aequifasciatus* Pellegrin, 1904
DISCUS BROWN „ALENQUER"
Brazil, wild

Photo: H.J. Mayland

S89426-4 *Symphysodon aequifasciatus* Pellegrin, 1904
DISCUS BROWN „ALENQUER"
Brazil, wild

Photo: Archiv A.C.S.

S89427-4 *Symphysodon aequifasciatus* Pellegrin, 1904
DISCUS BROWN „ALENQUER"
Brazil, wild

Photo: H.J. Mayland

S89428-4 *Symphysodon aequifasciatus* Pellegrin, 1904
DISCUS BROWN „ALENQUER"
Brazil, wild

Photo: W. Mikschofsky

S89429-4 *Symphysodon aequifasciatus* Pellegrin, 1904
DISCUS BROWN „ALENQUER"
Brazil, wild

Photo: W. Mikschofsky

S89430-4 *Symphysodon aequifasciatus* Pellegrin, 1904
DISCUS BROWN „ALENQUER"
Brazil, wild

Photo: H.J. Mayland

Gruppe 2 / Group 2: Discus Brown

S89431-4 *Symphysodon aequifasciatus* Pellegrin, 1904
DISCUS BROWN „ALENQUER"
Zuchtform / Breeding form

Photo: W. Konrad

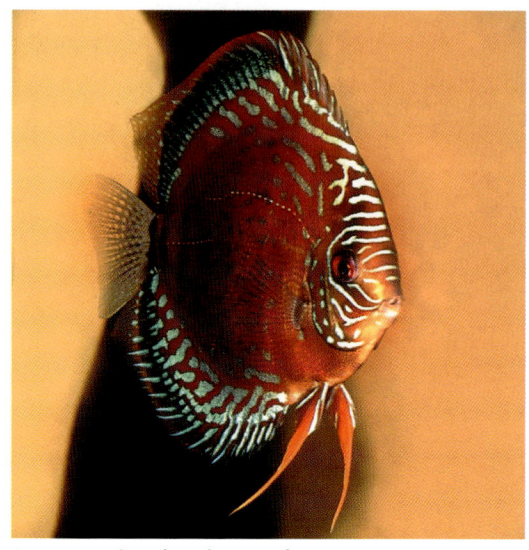

S89432-4 *Symphysodon aequifasciatus* Pellegrin, 1904
DISCUS BROWN „ALENQUER"
Zuchtform / Breeding form

Photo: W. Konrad

S89433-4 *Symphysodon aequifasciatus* Pellegrin, 1904
DISCUS BROWN „ALENQUER"
Zuchtform / Breeding form

Photo: W. Konrad

S89434-4 *Symphysodon aequifasciatus* Pellegrin, 1904
DISCUS BROWN „ALENQUER"
Zuchtform / Breeding form

Photo: Schlingmann

S89435-4 *Symphysodon aequifasciatus* Pellegrin, 1904
DISCUS BROWN „ALENQUER"
Zuchtform / Breeding form

Photo: W. Mikschofsky

S89436-4 *Symphysodon aequifasciatus* Pellegrin, 1904
DISCUS BROWN „ALENQUER"
Zuchtform / Breeding form

Photo: Schlingmann

Gruppe 2 / Group 2: Discus Brown

S89437-4 *Symphysodon aequifasciatus* Pellegrin, 1904
DISCUS BROWN „ALENQUER"
Brazil, wild

♂ Photo: H.J. Mayland

S89438-4 *Symphysodon aequifasciatus* Pellegrin, 1904
DISCUS BROWN „ALENQUER"
Brazil, wild

♀ Photo: H.J. Mayland

S90201-4 *Symphysodon aequifasciatus* Pellegrin, 1904
DISCUS BROWN „ALENQUER / RED EDDY"
Zuchtform / Breeding form F_1

♂ Photo: W. Konrad

S90202-4 *Symphysodon aequifasciatus* Pellegrin, 1904
DISCUS BROWN „ALENQUER / RED EDDY"
Zuchtform / Breeding form F_1

♀ Photo: W. Konrad

S90203-4 *Symphysodon aequifasciatus* Pellegrin, 1904
DISCUS BROWN „ALENQUER / RED EDDY"
Zuchtform / Breeding form F_2

♂ Photo: M. Göbel

S90204-4 *Symphysodon aequifasciatus* Pellegrin, 1904
DISCUS BROWN „ALENQUER / RED EDDY"
Zuchtform / Breeding form F_2

♀ Photo: W. Konrad

Gruppe 2 / Group 2: Discus Brown

S90205-4 *Symphysodon aequifasciatus* Pellegrin, 1904
DISCUS BROWN „ALENQUER / RED EDDY"
Zuchtform / Breeding form F_2

Photo: W. Konrad

S90206-4 *Symphysodon aequifasciatus* Pellegrin, 1904
DISCUS BROWN „ALENQUER / RED EDDY"
Zuchtform / Breeding form F_2

Photo: W. Konrad

S90207-4 *Symphysodon aequifasciatus* Pellegrin, 1904
DISCUS BROWN „ALENQUER / RED EDDY"
Zuchtform / Breeding form F_3

Photo: W. Konrad

S90208-4 *Symphysodon aequifasciatus* Pellegrin, 1904
DISCUS BROWN „ALENQUER / RED EDDY"
Zuchtform / Breeding form F_3

Photo: W. Konrad

S90209-4 *Symphysodon aequifasciatus* Pellegrin, 1904
DISCUS BROWN „ALENQUER / RED EDDY"
Zuchtform / Breeding form F_3

Photo: M. Göbel

S90210-4 *Symphysodon aequifasciatus* Pellegrin, 1904
DISCUS BROWN „ALENQUER / RED EDDY"
Zuchtform / Breeding form F_3

Photo: W. Konrad

South American **Cichlids** *IV* © Verlag A.C.S. GmbH

S90204-4 *Symphysodon aequifasciatus* Pellegrin, 1904
DISCUS BROWN "ALENQUER / RED EDDY"
Zuchtform / Breeding form F_2

Photo: W. Konrad

S90210-4 *Symphysodon aequifasciatus* Pellegrin, 1904
DISCUS BROWN „ALENQUER / RED EDDY"
Zuchtform / Breeding form F$_3$

Photo: W. Konrad

S90211-4 *Symphysodon aequifasciatus* Pellegrin, 1904
DISCUS BROWN „ALENQUER / RED EDDY"
Zuchtform / Breeding form F_4

Photo: M. Göbel

S90212-4 *Symphysodon aequifasciatus* Pellegrin, 1904
DISCUS BROWN „ALENQUER / RED EDDY"
Zuchtform / Breeding form F_4

Photo: M. Göbel

Gruppe 2 / Group 2: Discus Brown

S90213-4 *Symphysodon aequifasciatus* Pellegrin, 1904
DISCUS BROWN „ALENQUER / RED EDDY"
Zuchtform / Breeding form F_4

Photo: M. Göbel

South American Cichlids IV

S90211-4 *Symphysodon aequifascatus* Pellegrin, 1904
S90212-4 DISCUS ALENQUER "RED EDDY" PAAR / PAIR
Zuchtform / Breeding form F_4

Photo: M. Göbel

S89367-4 *Symphysodon aequifasciatus* Pellegrin, 1904
DISCUS BRAUN/BROWN
Brazil, wild

Photo: H.J. Mayland

S89368-4 *Symphysodon aequifasciatus* Pellegrin, 1904
DISCUS BRAUN/BROWN
Brazil, wild

Photo: W. Mikschofsky

S89369-4 *Symphysodon aequifasciatus* Pellegrin, 1904
DISCUS BRAUN/BROWN
Brazil, wild

Photo: W. Mikschofsky

S89370-4 *Symphysodon aequifasciatus* Pellegrin, 1904
DISCUS BRAUN/BROWN
Brazil, wild

Photo: Schlingmann

S89371-4 *Symphysodon aequifasciatus* Pellegrin, 1904
DISCUS BRAUN/BROWN
Brazil, wild

Photo: W. Mikschofsky

S89372-4 *Symphysodon aequifasciatus* Pellegrin, 1904
DISCUS BRAUN/BROWN
Brazil, wild

Photo: M. Göbel

Gruppe 2 / Group 2: Discus Brown

S89373-4 *Symphysodon aequifasciatus* Pellegrin, 1904
DISCUS BRAUN/BROWN
Brazil, wild

Photo: M. Göbeld

S89374-4 *Symphysodon aequifasciatus* Pellegrin, 1904
DISCUS BRAUN/BROWN
Brazil, wild

Photo: H.J. Mayland

S89375-4 *Symphysodon aequifasciatus* Pellegrin, 1904
DISCUS BRAUN/BROWN
Brazil, wild

Photo: Nakano / Archiv A.C.S.

S89376-4 *Symphysodon aequifasciatus* Pellegrin, 1904
DISCUS BRAUN/BROWN
Brazil, wild

Photo: Nakano / Archiv A.C.S.

S89377-4 *Symphysodon aequifasciatus* Pellegrin, 1904
DISCUS BRAUN/BROWN
Brazil, wild

Photo: U. Werner

S89378-4 *Symphysodon aequifasciatus* Pellegrin, 1904
DISCUS BRAUN/BROWN
Brazil, wild

Photo: U. Werner

S89379-4 *Symphysodon aequifasciatus* Pellegrin, 1904
DISCUS BRAUN/BROWN
Brazil, wild

Photo: Nakano / Archiv A.C.S.

S89380-4 *Symphysodon aequifasciatus* Pellegrin, 1904
DISCUS BRAUN/BROWN
Brazil, wild / Rio Purus

Photo: U. Werner

S89381-4 *Symphysodon aequifasciatus* Pellegrin, 1904
DISCUS BRAUN/BROWN
Brazil, wild / Rio Purus

Photo: U. Werner

S89382-4 *Symphysodon aequifasciatus* Pellegrin, 1904
DISCUS BRAUN/BROWN
Brazil, wild

Photo: H.J. Mayland

S89383-4 *Symphysodon aequifasciatus* Pellegrin, 1904
DISCUS BRAUN/BROWN
Brazil, wild / Rio Ipixuna

Photo: U. Werner

S89384-4 *Symphysodon aequifasciatus* Pellegrin, 1904
DISCUS BRAUN/BROWN
Brazil, wild / Rio Ipixuna

Photo: U. Werner

Gruppe 2 / Group 2: Discus Brown

S89385-4 *Symphysodon aequifasciatus* Pellegrin, 1904
DISCUS BRAUN/BROWN
Brazil, wild

Photo: H.J. Mayland

S89386-4 *Symphysodon aequifasciatus* Pellegrin, 1904
DISCUS BRAUN/BROWN
Brazil, wild

Photo: H.J. Mayland

S89387-3 *Symphysodon aequifasciatus* Pellegrin, 1904
DISCUS BRAUN/BROWN
Brazil, wild

Photo: J. Stendker

S89388-4 *Symphysodon aequifasciatus* Pellegrin, 1904
DISCUS BRAUN/BROWN
Brazil, wild

Photo: J. Stendker

S89389-3 *Symphysodon aequifasciatus* Pellegrin, 1904
DISCUS BRAUN/BROWN
Brazil, wild

Photo: H.J. Mayland

S89390-2 *Symphysodon aequifasciatus* Pellegrin, 1904
DISCUS BRAUN/BROWN
Brazil, wild

Photo: Archiv A.C.S.

South American **Cichlids IV** © Verlag A.C.S. GmbH

S89501-4 *Symphysodon aequifasciatus* Pellegrin, 1904
DISCUS BLAU/BLUE
Brazil, wild

Photo: Nakano / Archiv A.C.S.

S89502-4 *Symphysodon aequifasciatus* Pellegrin, 1904
DISCUS BLAU/BLUE
Brazil, wild

Photo: Nakano / Archiv A.C.S.

S89503-4 *Symphysodon aequifasciatus* Pellegrin, 1904
DISCUS BLAU/BLUE
Brazil, wild

Photo: Nakano / Archiv A.C.S.

S89504-4 *Symphysodon aequifasciatus* Pellegrin, 1904
DISCUS BLAU/BLUE
Brazil, wild

Photo: Nakano / Archiv A.C.S.

S89505-4 *Symphysodon aequifasciatus* Pellegrin, 1904
DISCUS BLAU/BLUE
Brazil, wild

Photo: Nakano / Archiv A.C.S.

S89506-4 *Symphysodon aequifasciatus* Pellegrin, 1904
DISCUS BLAU/BLUE
Brazil, wild

Photo: Nakano / Archiv A.C.S.

Gruppe 3 / Group 3: Discus Blue

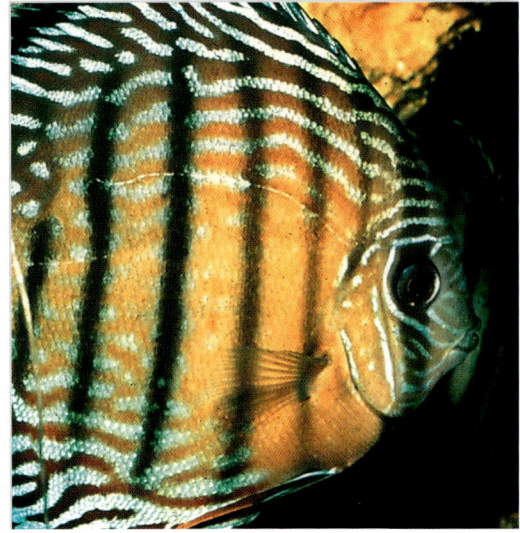

S89507-4 *Symphysodon aequifasciatus* Pellegrin, 1904
DISCUS BLAU/BLUE
Brazil, wild

Photo: U. Werner

S89508-4 *Symphysodon aequifasciatus* Pellegrin, 1904
DISCUS BLAU/BLUE
Brazil, wild

Photo: H.J. Mayland

S89509-4 *Symphysodon aequifasciatus* Pellegrin, 1904
DISCUS BLAU/BLUE
Brazil, wild

Photo: H.J. Mayland

S89510-4 *Symphysodon aequifasciatus* Pellegrin, 1904
DISCUS BLAU/BLUE
Brazil, wild

Photo: W. Konrad

S89511-4 *Symphysodon aequifasciatus* Pellegrin, 1904
DISCUS BLAU/BLUE
Brazil, wild

Photo: Nakano / Archiv A.C.S.

S89512-4 *Symphysodon aequifasciatus* Pellegrin, 1904
DISCUS BLAU/BLUE
Brazil, wild

Photo: Nakano / Archiv A.C.S.

S89513-4 *Symphysodon aequifasciatus* Pellegrin, 1904
DISCUS BLAU/BLUE
Brazil, wild

Photo: Nakano / Archiv A.C.S.

S89514-4 *Symphysodon aequifasciatus* Pellegrin, 1904
DISCUS BLAU/BLUE
Brazil, wild

Photo: Nakano / Archiv A.C.S.

S89515-4 *Symphysodon aequifasciatus* Pellegrin, 1904
DISCUS BLAU/BLUE
Brazil, wild, Rio Nhamunda(?)

Photo: U. Werner

S89516-4 *Symphysodon aequifasciatus* Pellegrin, 1904
DISCUS BLAU/BLUE
Brazil, wild, Rio Nhamunda(?)

Photo: U. Werner

S89517-4 *Symphysodon aequifasciatus* Pellegrin, 1904
DISCUS BLAU/BLUE
Brazil, wild, Rio Nhamunda(?)

Photo: U. Werner

S89518-4 *Symphysodon aequifasciatus* Pellegrin, 1904
DISCUS BLAU/BLUE
Brazil, wild, Rio Nhamunda(?)

Photo: U. Werner

Gruppe 3 / Group 3: Discus Blue

S89519-4 *Symphysodon aequifasciatus* Pellegrin, 1904
DISCUS BLAU/BLUE
Brazil, wild
Photo: H.J. Mayland

S89520-4 *Symphysodon aequifasciatus* Pellegrin, 1904
DISCUS BLAU/BLUE
Brazil, wild
Photo: H.J. Mayland

S89521-4 *Symphysodon aequifasciatus* Pellegrin, 1904
DISCUS BLAU/BLUE
Brazil, wild
Photo: H.J. Mayland

S89522-4 *Symphysodon aequifasciatus* Pellegrin, 1904
DISCUS BLAU/BLUE
Brazil, wild
Photo: Nakano / Archiv A.C.S.

S89523-4 *Symphysodon aequifasciatus* Pellegrin, 1904
DISCUS BLAU/BLUE
Brazil, wild
Photo: H.J. Mayland

S89524-4 *Symphysodon aequifasciatus* Pellegrin, 1904
DISCUS BLAU/BLUE
Brazil, wild Manacapurú(?)
Photo: H.J. Mayland

S89655-4 *Symphysodon aequifasciatus* PELLEGRIN, 1904
DISCUS BLAU/BLUE „MANACAPURÚ"
Zuchtform / Breeding form F_1

Photo: W. Konrad

S89656-4 *Symphysodon aequifasciatus* PELLEGRIN, 1904
DISCUS BLAU/BLUE „MANACAPURÚ"
Zuchtform / Breeding form F_1

Photo: W. Konrad

S89657-4 *Symphysodon aequifasciatus* PELLEGRIN, 1904
DISCUS BLAU/BLUE „MANACAPURÚ"
Zuchtform / Breeding form F_2

Photo: W. Konrad

S89658-4 *Symphysodon aequifasciatus* PELLEGRIN, 1904
DISCUS BLAU/BLUE „MANACAPURÚ"
Zuchtform / Breeding form F_2

Photo: W. Konrad

S89659-4 *Symphysodon aequifasciatus* PELLEGRIN, 1904
DISCUS BLAU/BLUE „MANACAPURÚ"
Zuchtform / Breeding form F_2

Photo: W. Konrad

S89659-4 *Symphysodon aequifasciatus* PELLEGRIN, 1904
DISCUS BLAU/BLUE „MANACAPURÚ"
Zuchtform / Breeding form F_2

Photo: W. Konrad

S89523-4 *Symphysodon aequifasciatus* Pellegrin, 1904
DISCUS BLAU/BLUE
Brazil, wild

Photo: H.J. Mayland

S89601-4 *Symphysodon aequifasciatus* Pellegrin, 1904
DISCUS „ROYAL BLUE"
Brazil, wild

Photo: F. Bodenmüller

S89602-4 *Symphysodon aequifasciatus* Pellegrin, 1904
DISCUS „ROYAL BLUE"
Brazil, wild

Photo: F. Bodenmüller

S89603-4 *Symphysodon aequifasciatus* Pellegrin, 1904
DISCUS „ROYAL BLUE"
Brazil, wild

Photo: Archiv A.C.S.

S89604-4 *Symphysodon aequifasciatus* Pellegrin, 1904
DISCUS „ROYAL BLUE"
Brazil, wild

Photo: Archiv A.C.S.

S89605-4 *Symphysodon aequifasciatus* Pellegrin, 1904
DISCUS „ROYAL BLUE"
Brazil, wild

Photo: Schlingmann

S89606-4 *Symphysodon aequifasciatus* Pellegrin, 1904
DISCUS „ROYAL BLUE"
Brazil, wild

Photo: Schlingmann

Gruppe 3 / Group 3: Discus Blue

South American Cichlids IV

S89610-4 *Symphysodon aequifasciatus* Pellegrin, 1904
DISCUS „ROYAL BLUE"   Männchen/Male
Brazil, wild

Photo: J. Schütz

S89611-4 *Symphysodon aequifasciatus* Pellegrin, 1904
DISCUS „ROYAL BLUE"   Weibchen/Female
Brazil, wild

Photo: J. Schütz

Gruppe 3 / Group 3: Discus Blue

S89624-4 *Symphysodon aequifasciatus* Pellegrin, 1904
DISCUS „ROYAL BLUE", F_2 von Tieren aus der Gruppe F_1 der Seiten 81 und 82
Photo: J. Schütz

S89625-4 *Symphysodon aequi...* DISCUS „ROYAL BLU... hes F_1 pp. 81 and 8...

S89612-4 *Symphysodon aequifasciatus* Pellegrin, 1904
DISCUS „ROYAL BLUE", F_1, alle Tiere auf S. 81 und 82 sind Geschwister; Eltern: s. S. 80
Photo: J. Schütz

S89613-4 *Symphysodon aequifasciatus* Pellegrin, 1904
DISCUS „ROYAL BLUE", F_1, all specimen on pp. 81 and 82 are brothers and sisters; parents: see p. 80
Photo: J. Schütz

S89626-4 *Symphysodon aequifasciatus* Pellegrin, 1904
DISCUS „ROYAL BLUE",
F_2 von Tieren aus der Gruppe F_1 der Seiten 81 und 82 / F_2 offspring from the fishes F_1 pp. 8...

S89614-4 *Symphysodon aequifasciatus* Pellegrin, 1904
DISCUS „ROYAL BLUE", F_1, alle Tiere auf S. 81 und 82 sind Geschwister; Eltern: s. S. 80
Photo: J. Schütz

S89615-4 *Symphysodon aequifasciatus* Pellegrin, 1904
DISCUS „ROYAL BLUE", F_1, all specimen on pp. 81 and 82 are brothers and sisters; parents: see p. 80
Photo: J. Schütz

S89616-4 *Symphysodon aequifasciatus* Pellegrin, 1904
DISCUS „ROYAL BLUE", F_1, alle Tiere auf S. 81 und 82 sind Geschwister; Eltern: s. S. 80
Photo: J. Schütz

S89617-4 *Symphysodon aequifasciatus* Pellegrin, 1904
DISCUS „ROYAL BLUE", F_1, all specimen on pp. 81 and 82 are brothers and sisters; parents: see p. 80
Photo: J. Schütz

Gruppe 3 / Group 3: Discus Blue

S89618-4 *Symphysodon aequifasciatus* Pellegrin, 1904
DISCUS „ROYAL BLUE", F_1, alle Tiere auf S. 81 und 82 sind Geschwister;   Eltern: s. S. 80

Photo: J. Schütz

S89620-4 *Symphysodon aequifasciatus* Pellegrin, 1904
DISCUS „ROYAL BLUE", F_1, alle Tiere auf S. 81 und 82 sind Geschwister;   Eltern: s. S. 80

Photo: J. Schütz

S89622-4 *Symphysodon aequifasciatus* Pellegrin, 1904
DISCUS „ROYAL BLUE", F_1, alle Tiere auf S. 81 und 82 sind Geschwister;   Eltern: s. S. 80

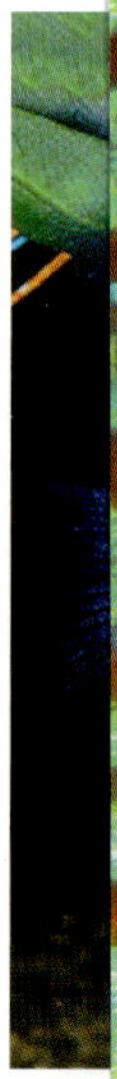

Photo: J. Schütz

S89659-4 *Symphysodon aequifasciatus* Pellegrin, 1904
DISCUS BLAU/BLUE „MANACAPURÚ"
Zuchtform / Breeding form F_2

Photo: W. Konrad

S89624-4 *Symphysodon aequifasciatus* Pellegrin, 1904
DISCUS „ROYAL BLUE", F$_2$ von Tieren aus der Gruppe F$_1$ der Seiten 81 und 82

Photo: J. Schütz

S89625-4 *Symphysodon aequifasciatus* Pellegrin, 1904
DISCUS „ROYAL BLUE", F2 offspring from the fishes F$_1$ pp. 81 and 82

Photo: J. Schütz

S89626-4 *Symphysodon aequifasciatus* Pellegrin, 1904
DISCUS „ROYAL BLUE",
F$_2$ von Tieren aus der Gruppe F$_1$ der Seiten 81 und 82 / F$_2$ offspring from the fishes F$_1$ pp. 81 and 82

Photo: J. Schütz

S89659-4 *Symphysodon aequifasciatus* Pellegrin, 1904
DISCUS BLAU/BLUE „MANACAPURÚ"
Zuchtform / Breeding form F_2

S89901-4 *Symphysodon aequifasciatus* Pellegrin, 1904
DISCUS GRÜN/GREEN
Brazil, wild

Photo: Nakano/Archiv A.C.S.

S89902-4 *Symphysodon aequifasciatus* Pellegrin, 1904
DISCUS GRÜN/GREEN
Brazil, wild

Photo: Nakano/Archiv A.C.S.

S89903-4 *Symphysodon aequifasciatus* Pellegrin, 1904
DISCUS GRÜN/GREEN
Brazil, wild

Photo: Nakano/Archiv A.C.S.

S89904-4 *Symphysodon aequifasciatus* Pellegrin, 1904
DISCUS GRÜN/GREEN
Brazil, wild

Photo: Nakano/Archiv A.C.S.

S89905-4 *Symphysodon aequifasciatus* Pellegrin, 1904
DISCUS GRÜN/GREEN
Brazil, wild

Photo: Nakano/Archiv A.C.S.

S89906-4 *Symphysodon aequifasciatus* Pellegrin, 1904
DISCUS GRÜN/GREEN
Brazil, wild

Photo: Nakano/Archiv A.C.S.

Gruppe 4 / Group 4: Discus Green

S89907-4 *Symphysodon aequifasciatus* Pellegrin, 1904
DISCUS GRÜN/GREEN
Brazil, wild   Rio Jurua(?)

Photo: U. Werner

S89908-4 *Symphysodon aequifasciatus* Pellegrin, 1904
DISCUS GRÜN/GREEN
Brazil, wild

Photo: F. Bodenmüller

S89909-4 *Symphysodon aequifasciatus* Pellegrin, 1904
DISCUS GRÜN/GREEN
Brazil, wild

Photo: H.J. Mayland

S89910-4 *Symphysodon aequifasciatus* Pellegrin, 1904
DISCUS GRÜN/GREEN
Brazil, wild   Rio Jurua(?)

Photo: U. Werner

S89911-4 *Symphysodon aequifasciatus* Pellegrin, 1904
DISCUS GRÜN/GREEN
Brazil, wild

Photo: H.J. Mayland

S89912-4 *Symphysodon aequifasciatus* Pellegrin, 1904
DISCUS GRÜN/GREEN
Brazil, wild

Photo: H.J. Mayland

S89913-4 *Symphysodon aequifasciatus* Pellegrin, 1904
DISCUS GRÜN/GREEN
Brazil, wild

Photo: Nakano/Archiv A.C.S.

S89914-4 *Symphysodon aequifasciatus* Pellegrin, 1904
DISCUS GRÜN/GREEN
Brazil, wild

Photo: Nakano/Archiv A.C.S.

S89953-4 *Symphysodon aequifasciatus* Pellegrin, 1904
DISCUS GRÜN/GREEN "TEFÉ"
Brazil, wild

Photo: U. Werner

S89954-4 *Symphysodon aequifasciatus* Pellegrin, 1904
DISCUS GRÜN/GREEN "TEFÉ"
Brazil, wild

Photo: U. Werner

S89956-4 *Symphysodon aequifasciatus* Pellegrin, 1904
DISCUS GRÜN/GREEN "TEFÉ"
Brazil, wild

Photo: F. Bodenmüller

S89964-4 *Symphysodon aequifasciatus* Pellegrin, 1904
DISCUS GRÜN/GREEN "TEFÉ"
Brazil, wild

Photo: U. Werner

Gruppe 4 / Group 4: Discus Green

S89915-4 *Symphysodon aequifasciatus* PELLEGRIN, 1904
DISCUS GRÜN/GREEN
Brazil, wild

Photo: Schlingmann

S89916-4 *Symphysodon aequifasciatus* PELLEGRIN, 1904
DISCUS GRÜN/GREEN
Brazil, wild

Photo: Schlingmann

S89917-4 *Symphysodon aequifasciatus* PELLEGRIN, 1904
DISCUS GRÜN/GREEN
Brazil, wild

Photo: W. Mikschofsky

S89918-4 *Symphysodon aequifasciatus* PELLEGRIN, 1904
DISCUS GRÜN/GREEN
Brazil, wild

Photo: H.J. Mayland

S89919-4 *Symphysodon aequifasciatus* PELLEGRIN, 1904
DISCUS GRÜN/GREEN
Brazil, wild

Photo: M. Göbel

S89920-4 *Symphysodon aequifasciatus* PELLEGRIN, 1904
DISCUS GRÜN/GREEN
Brazil, wild

Photo: M. Göbel

South American **Cichlids** IV © Verlag A.C.S. GmbH

S89921-4 *Symphysodon aequifasciatus* Pellegrin, 1904
DISCUS GRÜN/GREEN
Brazil, wild

Photo: H.J. Mayland

S89922-4 *Symphysodon aequifasciatus* Pellegrin, 1904
DISCUS GRÜN/GREEN
Brazil, wild

Photo: W. Mikschofsky

S89923-4 *Symphysodon aequifasciatus* Pellegrin, 1904
DISCUS GRÜN/GREEN
Brazil, wild

Photo: M. Göbel

S89924-4 *Symphysodon aequifasciatus* Pellegrin, 1904
DISCUS GRÜN/GREEN
Brazil, wild

Photo: W. Mikschofsky

S89925-4 *Symphysodon aequifasciatus* Pellegrin, 1904
DISCUS GRÜN/GREEN
Brazil, wild

Photo: W. Mikschofsky

S89926-4 *Symphysodon aequifasciatus* Pellegrin, 1904
DISCUS GRÜN/GREEN
Brazil, wild

Photo: H.J. Mayland

Gruppe 4 / Group 4: Discus Green

S89927-4 *Symphysodon aequifasciatus* Pellegrin, 1904
DISCUS GRÜN/GREEN
Brazil, wild

Photo: H.J. Mayland

S89928-4 *Symphysodon aequifasciatus* Pellegrin, 1904
DISCUS GRÜN/GREEN
Brazil, wild

Photo: H.J. Mayland

S89929-4 *Symphysodon aequifasciatus* Pellegrin, 1904
DISCUS GRÜN/GREEN
Brazil, wild

Photo: H.J. Mayland

S89930-4 *Symphysodon aequifasciatus* Pellegrin, 1904
DISCUS GRÜN/GREEN
Brazil, wild

Photo: H.J. Mayland

S89931-4 *Symphysodon aequifasciatus* Pellegrin, 1904
DISCUS GRÜN/GREEN
Brazil, wild

Photo: H.J. Mayland

S89932-4 *Symphysodon aequifasciatus* Pellegrin, 1904
DISCUS GRÜN/GREEN
Brazil, wild

Photo: H.J. Mayland

S89933-4 *Symphysodon aequifasciatus* PELLEGRIN, 1904
DISCUS GRÜN/GREEN
Brazil, wild

Photo: W. Mikschofsky

S89934-4 *Symphysodon aequifasciatus* PELLEGRIN, 1904
DISCUS GRÜN/GREEN
Brazil, wild

Photo: Archiv A.C.S.

S89935-4 *Symphysodon aequifasciatus* PELLEGRIN, 1904
DISCUS GRÜN/GREEN
Brazil, wild

Photo: M. Göbel

S89936-4 *Symphysodon aequifasciatus* PELLEGRIN, 1904
DISCUS GRÜN/GREEN
Brazil, wild

Photo: M. Göbel

S89937-4 *Symphysodon aequifasciatus* PELLEGRIN, 1904
DISCUS GRÜN/GREEN
Brazil, wild

Photo: W. Konrad

S89938-4 *Symphysodon aequifasciatus* PELLEGRIN, 1904
DISCUS GRÜN/GREEN
Brazil, wild

Photo: W. Konrad

Gruppe 4 / Group 4: Discus Green

S89939-4 *Symphysodon aequifasciatus* Pellegrin, 1904
DISCUS GRÜN/GREEN
Zuchtform / Breeding form, F_1

Photo: M. Göbel

S89940-4 *Symphysodon aequifasciatus* Pellegrin, 1904
DISCUS GRÜN/GREEN
Brazil, wild

Photo: F. Bodenmüller

S89941-4 *Symphysodon aequifasciatus* Pellegrin, 1904
DISCUS GRÜN/GREEN
Brazil, wild

Photo: F. Bodenmüller

S89942-4 *Symphysodon aequifasciatus* Pellegrin, 1904
DISCUS GRÜN/GREEN
Brazil, wild   Rio Putumajo(?)

Photo: U. Werner

S89943-4 *Symphysodon aequifasciatus* Pellegrin, 1904
DISCUS GRÜN/GREEN
Brazil, wild

Photo: H.J. Mayland

S89944-4 *Symphysodon aequifasciatus* Pellegrin, 1904
DISCUS GRÜN/GREEN
Brazil, wild

Photo: M. Göbel

S89945-4 *Symphysodon aequifasciatus* PELLEGRIN, 1904
DISCUS GRÜN/GREEN
Brazil, wild

Photo: H.J. Mayland

S89947-4 *Symphysodon aequifasciatus* PELLEGRIN, 1904
DISCUS GRÜN/GREEN
Brazil, wild

Photo: W. Mikschofsky

S89948-4 *Symphysodon aequifasciatus* PELLEGRIN, 1904
DISCUS GRÜN/GREEN "ROYAL-GREEN"
Brazil, wild

Photo: F. Bodenmüller

Gruppe 4 / Group 4: Discus Green

South American Cichlids IV

S89944-4 *Symphysodon aequifasciatus* Pellegrin, 1904
DISCUS GRÜN/GREEN
Brazil, wild

Photo: M. Göbel

S89951-4 *Symphysodon aequifasciatus* Pellegrin, 1904
DISCUS GRÜN/GREEN
Brazil, wild

Photo: F. Bodenmüller

S89952-4 *Symphysodon aequifasciatus* Pellegrin, 1904
DISCUS GRÜN/GREEN
Brazil, wild

Photo: F. Bodenmüller

S89955-4 *Symphysodon aequifasciatus* Pellegrin, 1904
DISCUS GRÜN/GREEN
Brazil, wild

Photo: F. Bodenmüller

South American **Cichlids IV**

Gruppe 4 / Group 4: Discus Green

S89957-4 *Symphysodon aequifasciatus* Pellegrin, 1904
DISCUS GRÜN/GREEN
Brazil, wild

Photo: W.A. Tomey

S89958-4 *Symphysodon aequifasciatus* Pellegrin, 1904
DISCUS GRÜN/GREEN
Brazil, wild

Photo: W.A. Tomey

S89959-4 *Symphysodon aequifasciatus* Pellegrin, 1904
DISCUS GRÜN/GREEN "SUNSET-RED"
Brazil, wild

Photo: Wayne DC HongKong

S89960-4 *Symphysodon aequifasciatus* Pellegrin, 1904
DISCUS GRÜN/GREEN "NOBEL-RED"
Brazil, wild

Photo: Wayne DC HongKong

S89961-4 *Symphysodon aequifasciatus* Pellegrin, 1904
DISCUS GRÜN/GREEN
Brazil, wild

Photo: Archiv A.C.S.

S89962-4 *Symphysodon aequifasciatus* Pellegrin, 1904
DISCUS GRÜN/GREEN
Brazil, wild

Photo: M. Göbel

Gruppe 4 / Group 4: Discus Green

S89963-4 *Symphysodon aequifasciatus* Pellegrin, 1904
DISCUS GRÜN/GREEN "COARI"
Brazil, wild

Photo: W. Konrad

S91501-4 *Symphysodon aequifasciatus* Pellegrin, 1904
DISCUS GRÜN/GREEN "COARI"
Zuchtform / Breeding form F_1 from S89963

Photo: W. Konrad

S89965-4 *Symphysodon aequifasciatus* Pellegrin, 1904
DISCUS GRÜN/GREEN "COARI"
Brazil, wild

Photo: M. Göbel

S89966-4 *Symphysodon aequifasciatus* Pellegrin, 1904
DISCUS GRÜN/GREEN "COARI"
Brazil, wild

Photo: M. Göbel

S89967-4 *Symphysodon aequifasciatus* Pellegrin, 1904
DISCUS GRÜN/GREEN "COARI"
Brazil, wild

Photo: M. Göbel

S91504-4 *Symphysodon aequifasciatus* Pellegrin, 1904
DISCUS GRÜN/GREEN "COARI"
Zuchtform / Breeding form F_1

Photo: M. Göbel

South American Cichlids IV

Gruppe 4 / Group 4: Discus Green

S89969-4 *Symphysodon aequifasciatus* Pellegrin, 1904
DISCUS GRÜN/GREEN "COARI"
Brazil, wild

Photo: M. Göbel

S91502-4 *Symphysodon aequifasciatus* Pellegrin, 1904
DISCUS GRÜN/GREEN "COARI"
Zuchtform / Breeding form F_1

Photo: M. Göbel

S91503-4 *Symphysodon aequifasciatus* Pellegrin, 1904
DISCUS GRÜN/GREEN "COARI"
Zuchtform / Breeding form F_1

Photo: M. Göbel

S91505-4 *Symphysodon aequifasciatus* Pellegrin, 1904
DISCUS GRÜN/GREEN "RED SPOTTED"
Zuchtform / Breeding form

Photo: M. Göbel

S91506-3 *Symphysodon aequifasciatus* Pellegrin, 1904
DISCUS GRÜN/GREEN "RED SPOTTED"
Zuchtform / Breeding form

Photo: M. Göbel

S91507-4 *Symphysodon aequifasciatus* Pellegrin, 1904
DISCUS GRÜN/GREEN "RED SPOTTED"
Zuchtform / Breeding form

Photo: M. Göbel

S91508-4 *Symphysodon aequifasciatus* Pellegrin, 1904
DISCUS GRÜN/GREEN "RED SPOTTED"
Zuchtform / Breeding form

Photo: M. Göbel

S91509-4 *Symphysodon aequifasciatus* Pellegrin, 1904
DISCUS GRÜN/GREEN "RED SPOTTED"
Zuchtform / Breeding form

Photo: M. Göbel

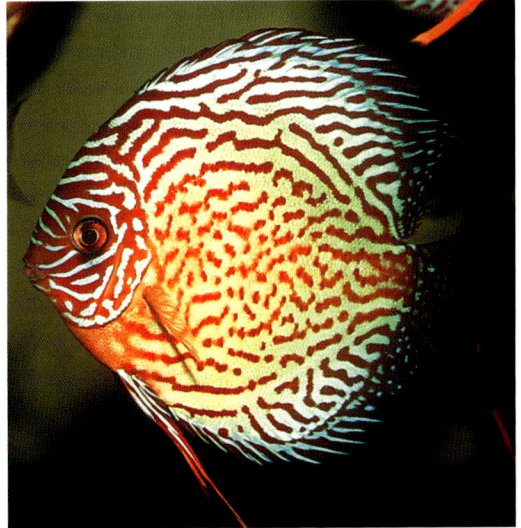

S91510-4 *Symphysodon aequifasciatus* Pellegrin, 1904
DISCUS GRÜN/GREEN "RED SPOTTED"
Zuchtform / Breeding form

Photo: M. Göbel

Gruppe 4 / Group 4: Discus Green

S91511-4 *Symphysodon aequifasciatus* Pellegrin, 1904
DISCUS GRÜN/GREEN "RED SPOTTED"
Zuchtform / Breeding form

Photo: M. Göbel

S91512-4 *Symphysodon aequifasciatus* Pellegrin, 1904
DISCUS GRÜN/GREEN "RED SPOTTED"
Zuchtform / Breeding form

Photo: M. Göbel

S91513-4 *Symphysodon aequifasciatus* Pellegrin, 1904
DISCUS GRÜN/GREEN "RED SPOTTED"
Zuchtform / Breeding form

Photo: M. Göbel

S91514-4 *Symphysodon aequifasciatus* Pellegrin, 1904
DISCUS GRÜN/GREEN "RED SPOTTED"
Zuchtform / Breeding form

Photo: M. Göbel

S91515-4 *Symphysodon aequifasciatus* Pellegrin, 1904
DISCUS GRÜN/GREEN "RED SPOTTED"
Zuchtform / Breeding form

Photo: M. Göbel

S91516-4 *Symphysodon aequifasciatus* Pellegrin, 1904
DISCUS GRÜN/GREEN "RED SPOTTED"
Zuchtform / Breeding form

Photo: M. Göbel

Gruppe 4 / Group 4: Discus Green

S91520-5 *Symphysodon aequifasciatus* Pellegrin, 1904
DISCUS GRÜN/GREEN "RED SPOTTED"
Zuchtform / Breeding form F$_5$

Photo: J. Schütz

S91521-5 *Symphysodon aequifasciatus* Pellegrin, 1904
DISCUS GRÜN/GREEN "RED SPOTTED"
Zuchtform / Breeding form F$_5$

Photo: J. Schütz

S91522-5 *Symphysodon aequifasciatus* Pellegrin, 1904
DISCUS GRÜN/GREEN "RED SPOTTED"
Zuchtform / Breeding form F$_5$

Photo: J. Schütz

S91523-5 *Symphysodon aequifasciatus* Pellegrin, 1904
DISCUS GRÜN/GREEN "RED SPOTTED"
Zuchtform / Breeding form F$_5$

Photo: J. Schütz

S91524-5 *Symphysodon aequifasciatus* Pellegrin, 1904
DISCUS GRÜN/GREEN "RED SPOTTED"
Zuchtform / Breeding form F$_5$

Photo: J. Schütz

S91525-5 *Symphysodon aequifasciatus* Pellegrin, 1904
DISCUS GRÜN/GREEN "RED SPOTTED"
Zuchtform / Breeding form F$_5$

Photo: J. Schütz

South American Cichlids IV

Gruppe 4 / Group 4: Discus Green

S91527-5 *Symphysodon aequifasciatus* Pellegrin, 1904
DISCUS GRÜN/GREEN "RED SPOTTED"
Zuchtform / Breeding form F_5

Photo: M. Göbel

S91527-5 *Symphysodon aequifasciatus* Pellegrin, 1904
DISCUS GRÜN/GREEN "RED SPOTTED"
Zuchtform / Breeding form F_5

Photo: M. Göbel

S91528-5 *Symphysodon aequifasciatus* Pellegrin, 1904
DISCUS GRÜN/GREEN "RED SPOTTED"
Zuchtform / Breeding form F_5

Photo: M. Göbel

S91529-5 *Symphysodon aequifasciatus* Pellegrin, 1904
DISCUS GRÜN/GREEN "RED SPOTTED"
Zuchtform / Breeding form F_5

Photo: M. Göbel

S91530-5 *Symphysodon aequifasciatus* Pellegrin, 1904
DISCUS GRÜN/GREEN "RED SPOTTED"
Zuchtform / Breeding form F_5

Photo: M. Göbel

S91526-5 *Symphysodon aequifasciatus* Pellegrin, 1904
DISCUS GRÜN/GREEN "RED SPOTTED"
Zuchtform / Breeding form F_5

Photo: M. Göbel

S91532-3 *Symphysodon aequifasciatus* Pellegrin, 1904
DISCUS GRÜN/GREEN "RED SPOTTED"
Zuchtform / Breeding form F_5

Photo: J. Schütz

S91533-3 *Symphysodon aequifasciatus* Pellegrin, 1904
DISCUS GRÜN/GREEN "RED SPOTTED"
Zuchtform / Breeding form F_5

Photo: J. Schütz

S91535-5 *Symphysodon aequifasciatus* Pellegrin, 1904
DISCUS GRÜN/GREEN "RED SPOTTED"
Zuchtform / Breeding form F_5

Photo: M. Göbel

S73101-4 *Symphysodon aequifasciatus* Pellegrin, 1904
DISCUS TÜRKIS/TURQUOISE
Zuchtform / Breeding form (photo from 1968)

Photo: M. Göbel

S73102-4 *Symphysodon aequifasciatus* Pellegrin, 1904
DISCUS TÜRKIS/TURQUOISE
Zuchtform / Breeding form

Photo: H. Nakano

S73103-4 *Symphysodon aequifasciatus* Pellegrin, 1904
DISCUS TÜRKIS/TURQUOISE
Zuchtform / Breeding form

Photo: M. Göbel

S73104-4 *Symphysodon aequifasciatus* Pellegrin, 1904
DISCUS TÜRKIS/TURQUOISE
Zuchtform / Breeding form

Photo: H. Nakano

S73105-4 *Symphysodon aequifasciatus* Pellegrin, 1904
DISCUS TÜRKIS/TURQUOISE
Zuchtform / Breeding form

Photo: H. Nakano

S73106-4 *Symphysodon aequifasciatus* Pellegrin, 1904
DISCUS TÜRKIS/TURQUOISE
Zuchtform / Breeding form

Photo: W.A. Tomey

Gruppe 5 / Group 5: Türkis gezeichnete Diskus / Turquoise Striped Discus

Gruppe 5 / Group 5: Türkis gezeichnete Diskus / Turquoise Striped Discus

S73107-4 *Symphysodon aequifasciatus* PELLEGRIN, 1904
DISCUS TÜRKIS/TURQUOISE
Zuchtform / Breeding form

Photo: H.J. Mayland

S73108-4 *Symphysodon aequifasciatus* PELLEGRIN, 1904
DISCUS TÜRKIS/TURQUOISE
Zuchtform / Breeding form

Photo: H.J. Mayland

S73109-4 *Symphysodon aequifasciatus* PELLEGRIN, 1904
DISCUS TÜRKIS/TURQUOISE
Zuchtform / Breeding form

Photo: H.J. Mayland

S73110-4 *Symphysodon aequifasciatus* PELLEGRIN, 1904
DISCUS TÜRKIS/TURQUOISE
Zuchtform / Breeding form

Photo: H.J. Mayland

S73111-4 *Symphysodon aequifasciatus* PELLEGRIN, 1904
DISCUS TÜRKIS/TURQUOISE "COBALT"
Zuchtform / Breeding form

Photo: U. Werner

S73112-4 *Symphysodon aequifasciatus* PELLEGRIN, 1904
DISCUS TÜRKIS/TURQUOISE
Zuchtform / Breeding form

Photo: H.J. Mayland

S73113-4 *Symphysodon aequifasciatus* Pellegrin, 1904
DISCUS TÜRKIS/TURQUOISE
Zuchtform / Breeding form

Photo: H.J. Mayland

S73114-4 *Symphysodon aequifasciatus* Pellegrin, 1904
DISCUS TÜRKIS/TURQUOISE
Zuchtform / Breeding form

Photo: H.J. Mayland

S73115-4 *Symphysodon aequifasciatus* Pellegrin, 1904
DISCUS TÜRKIS/TURQUOISE
Zuchtform / Breeding form

Photo: H.J. Mayland

S73116-4 *Symphysodon aequifasciatus* Pellegrin, 1904
DISCUS TÜRKIS/TURQUOISE
Zuchtform / Breeding form

Photo: H.J. Mayland

S73117-4 *Symphysodon aequifasciatus* Pellegrin, 1904
DISCUS TÜRKIS/TURQUOISE
Zuchtform / Breeding form

Photo: H.J. Mayland

S73118-4 *Symphysodon aequifasciatus* Pellegrin, 1904
DISCUS TÜRKIS/TURQUOISE
Zuchtform / Breeding form

Photo: J. Stendker

Gruppe 5 / Group 5: Türkis gezeichnete Diskus / Turquoise Striped Discus

Gruppe 5 / Group 5: Türkis gezeichnete Diskus / Turquoise Striped Discus

S73119-4 *Symphysodon aequifasciatus* Pellegrin, 1904
DISCUS TÜRKIS/TURQUOISE
Zuchtform / Breeding form

Photo: H.J. Mayland

S73120-4 *Symphysodon aequifasciatus* Pellegrin, 1904
DISCUS TÜRKIS/TURQUOISE
Zuchtform / Breeding form

Photo: H.J. Mayland

S73121-4 *Symphysodon aequifasciatus* Pellegrin, 1904
DISCUS TÜRKIS/TURQUOISE
Zuchtform / Breeding form

Photo: H.J. Mayland

S73122-4 *Symphysodon aequifasciatus* Pellegrin, 1904
DISCUS TÜRKIS/TURQUOISE
Zuchtform / Breeding form

Photo: H.J. Mayland

S73123-4 *Symphysodon aequifasciatus* Pellegrin, 1904
DISCUS TÜRKIS/TURQUOISE
Zuchtform / Breeding form

Photo: Nakano / Archiv A.C.S.

S73124-4 *Symphysodon aequifasciatus* Pellegrin, 1904
DISCUS TÜRKIS/TURQUOISE
Zuchtform / Breeding form

Photo: ACS/TFF Mainland

S73125-4 *Symphysodon aequifasciatus* Pellegrin, 1904
DISCUS TÜRKIS/TURQUOISE
Zuchtform / Breeding form

Photo: M. Göbel

S73126-4 *Symphysodon aequifasciatus* Pellegrin, 1904
DISCUS TÜRKIS/TURQUOISE
Zuchtform / Breeding form

Photo: M. Göbel

S73127-4 *Symphysodon aequifasciatus* Pellegrin, 1904
DISCUS TÜRKIS/TURQUOISE
Zuchtform / Breeding form

Photo: M. Göbel

S73128-4 *Symphysodon aequifasciatus* Pellegrin, 1904
DISCUS TÜRKIS/TURQUOISE
Zuchtform / Breeding form

Photo: Schlingmann

S73129-4 *Symphysodon aequifasciatus* Pellegrin, 1904
DISCUS TÜRKIS/TURQUOISE
Zuchtform / Breeding form

Photo: M. Göbel

S73130-4 *Symphysodon aequifasciatus* Pellegrin, 1904
DISCUS TÜRKIS/TURQUOISE
Zuchtform / Breeding form

Photo: H.J. Mayland

Gruppe 5 / Group 5: Türkis gezeichnete Diskus / Turquoise Striped Discus

S73119-4 *Symphysodon aequifasciatus* Pellegrin, 1904
DISCUS TÜRKIS/TURQUOISE
Zuchtform / Breeding form

Photo: H.J. Mayland

Gruppe 5 / Group 5: Türkis gezeichnete Diskus / Turquoise Striped Discus

S73131-4 *Symphysodon aequifasciatus* Pellegrin, 1904
DISCUS TÜRKIS/TURQUOISE
Zuchtform / Breeding form
Photo: H.J. Mayland

S73132-4 *Symphysodon aequifasciatus* Pellegrin, 1904
DISCUS TÜRKIS/TURQUOISE
Zuchtform / Breeding form
Photo: H.J. Mayland

S73133-4 *Symphysodon aequifasciatus* Pellegrin, 1904
DISCUS TÜRKIS/TURQUOISE
Zuchtform / Breeding form
Photo: H.J. Mayland

S73134-4 *Symphysodon aequifasciatus* Pellegrin, 1904
DISCUS TÜRKIS/TURQUOISE
Zuchtform / Breeding form
Photo: M. Göbel

S73135-4 *Symphysodon aequifasciatus* Pellegrin, 1904
DISCUS TÜRKIS/TURQUOISE
Zuchtform / Breeding form
Photo: H.J. Mayland

S73136-4 *Symphysodon aequifasciatus* Pellegrin, 1904
DISCUS TÜRKIS/TURQUOISE
Zuchtform / Breeding form
Photo: H.J. Mayland

South American Cichlids IV

Gruppe 5 / Group 5: Türkis gezeichnete Diskus / Turquoise Striped Discus

S73137-4 *Symphysodon aequifasciatus* Pellegrin, 1904
DISCUS TÜRKIS/TURQUOISE
Zuchtform / Breeding form

Photo: H.J. Mayland

S73138-4 *Symphysodon aequifasciatus* Pellegrin, 1904
DISCUS TÜRKIS/TURQUOISE
Zuchtform / Breeding form

Photo: H.J. Mayland

S73139-4 *Symphysodon aequifasciatus* Pellegrin, 1904
DISCUS TÜRKIS/TURQUOISE
Zuchtform / Breeding form

Photo: H.J. Mayland

S73140-4 *Symphysodon aequifasciatus* Pellegrin, 1904
DISCUS TÜRKIS/TURQUOISE
Zuchtform / Breeding form

Photo: H.J. Mayland

S73141-4 *Symphysodon aequifasciatus* Pellegrin, 1904
DISCUS TÜRKIS/TURQUOISE
Zuchtform / Breeding form

Photo: Archiv A.C.S./Tomizana

S73142-4 *Symphysodon aequifasciatus* Pellegrin, 1904
DISCUS TÜRKIS/TURQUOISE
Zuchtform / Breeding form

Photo: Archiv A.C.S./Tomizana

S73143-4 *Symphysodon aequifasciatus* Pellegrin, 1904
DISCUS TÜRKIS/TURQUOISE
Zuchtform / Breeding form

Photo: M. Göbel

S73144-4 *Symphysodon aequifasciatus* Pellegrin, 1904
DISCUS TÜRKIS/TURQUOISE
Zuchtform / Breeding form

Photo: M. Göbel

S73145-4 *Symphysodon aequifasciatus* Pellegrin, 1904
DISCUS TÜRKIS/TURQUOISE
Zuchtform / Breeding form

Photo: M. Göbel

S73146-4 *Symphysodon aequifasciatus* Pellegrin, 1904
DISCUS TÜRKIS/TURQUOISE
Zuchtform / Breeding form

Photo: M. Göbel

S73147-4 *Symphysodon aequifasciatus* Pellegrin, 1904
DISCUS TÜRKIS/TURQUOISE
Zuchtform / Breeding form

Photo: M. Göbel

S73148-4 *Symphysodon aequifasciatus* Pellegrin, 1904
DISCUS TÜRKIS/TURQUOISE
Zuchtform / Breeding form

Photo: M. Göbel

Gruppe 5 / Group 5: Türkis gezeichnete Diskus / Turquoise Striped Discus

Gruppe 5 / Group 5: Türkis gezeichnete Diskus / Turquoise Striped Discus

S73149-4 *Symphysodon aequifasciatus* PELLEGRIN, 1904
DISCUS TÜRKIS/TURQUOISE
Zuchtform / Breeding form

Photo: M. Göbel

S73150-4 *Symphysodon aequifasciatus* PELLEGRIN, 1904
DISCUS TÜRKIS/TURQUOISE
Zuchtform / Breeding form

Photo: M. Göbel

S73151-4 *Symphysodon aequifasciatus* PELLEGRIN, 1904
DISCUS TÜRKIS/TURQUOISE
Zuchtform / Breeding form

Photo: M. Göbel

S73152-4 *Symphysodon aequifasciatus* PELLEGRIN, 1904
DISCUS TÜRKIS/TURQUOISE
Zuchtform / Breeding form

Photo: M. Göbel

S73153-4 *Symphysodon aequifasciatus* PELLEGRIN, 1904
DISCUS TÜRKIS/TURQUOISE
Zuchtform / Breeding form

Photo: M. Göbel

S73154-4 *Symphysodon aequifasciatus* PELLEGRIN, 1904
DISCUS TÜRKIS/TURQUOISE
Zuchtform / Breeding form

Photo: M. Göbel

S73144-4 *Symphysodon aequifasciatus* Pellegrin, 1904
DISCUS TÜRKIS/TURQUOISE
Zuchtform / Breeding form

S73154-4 *Symphysodon aequifasciatus* Pellegrin, 1904
DISCUS TÜRKIS/TURQUOISE
Zuchtform / Breeding form

S73155-4 *Symphysodon aequifasciatus* Pellegrin, 1904
DISCUS TÜRKIS/TURQUOISE
Zuchtform / Breeding form

Photo: H. Nakano

S73156-4 *Symphysodon aequifasciatus* Pellegrin, 1904
DISCUS TÜRKIS/TURQUOISE
Zuchtform / Breeding form

Photo: H. Nakano

S73157-4 *Symphysodon aequifasciatus* Pellegrin, 1904
DISCUS TÜRKIS/TURQUOISE
Zuchtform / Breeding form

Photo: M. Göbel

S73158-4 *Symphysodon aequifasciatus* Pellegrin, 1904
DISCUS TÜRKIS/TURQUOISE
Zuchtform / Breeding form

Photo: M. Göbel

S73159-4 *Symphysodon aequifasciatus* Pellegrin, 1904
DISCUS TÜRKIS/TURQUOISE
Zuchtform / Breeding form

Photo: M. Göbel

S73160-4 *Symphysodon aequifasciatus* Pellegrin, 1904
DISCUS TÜRKIS/TURQUOISE
Zuchtform / Breeding form

Photo: Tomizana / A.C.S.

Gruppe 5 / Group 5: Türkis gezeichnete Diskus / Turquoise Striped Discus

Gruppe 5 / Group 5: Türkis gezeichnete Diskus / Turquoise Striped Discus

S73161-4 *Symphysodon aequifasciatus* Pellegrin, 1904
DISCUS TÜRKIS/TURQUOISE
Zuchtform / Breeding form

Photo: Archiv A.C.S.

S73162-4 *Symphysodon aequifasciatus* Pellegrin, 1904
DISCUS TÜRKIS/TURQUOISE
Zuchtform / Breeding form

Photo: Archiv A.C.S.

S73163-4 *Symphysodon aequifasciatus* Pellegrin, 1904
DISCUS TÜRKIS/TURQUOISE
Zuchtform / Breeding form

Photo: H.J. Mayland

S73164-4 *Symphysodon aequifasciatus* Pellegrin, 1904
DISCUS TÜRKIS/TURQUOISE
Zuchtform / Breeding form

Photo: H. Nakano

S73165-4 *Symphysodon aequifasciatus* Pellegrin, 1904
DISCUS TÜRKIS/TURQUOISE
Zuchtform / Breeding form

Photo: M. Göbel

S73166-4 *Symphysodon aequifasciatus* Pellegrin, 1904
DISCUS TÜRKIS/TURQUOISE
Zuchtform / Breeding form

Photo: M. Göbel

S73173-4 *Symphysodon aequifasciatus* PELLEGRIN, 1904
DISCUS TÜRKIS/TURQUOISE
Zuchtform / Breeding form

Photo: J. Stendker

Gruppe 5 / Group 5: Türkis gezeichnete Diskus / Turquoise Striped Discus

S73167-4 *Symphysodon aequifasciatus* Pellegrin, 1904
DISCUS TÜRKIS/TURQUOISE
Zuchtform / Breeding form

Photo: J. Stendker

S73168-4 *Symphysodon aequifasciatus* Pellegrin, 1904
DISCUS TÜRKIS/TURQUOISE
Zuchtform / Breeding form

Photo: J. Stendker

S73169-4 *Symphysodon aequifasciatus* Pellegrin, 1904
DISCUS TÜRKIS/TURQUOISE
Zuchtform / Breeding form

Photo: J. Stendker

S73170-4 *Symphysodon aequifasciatus* Pellegrin, 1904
DISCUS TÜRKIS/TURQUOISE
Zuchtform / Breeding form

Photo: J. Stendker

S73171-4 *Symphysodon aequifasciatus* Pellegrin, 1904
DISCUS TÜRKIS/TURQUOISE
Zuchtform / Breeding form

Photo: J. Stendker

S73172-4 *Symphysodon aequifasciatus* Pellegrin, 1904
DISCUS TÜRKIS/TURQUOISE
Zuchtform / Breeding form

Photo: J. Stendker

S73179-4 *Symphysodon aequifasciatus* Pellegrin, 1904
DISCUS TÜRKIS/TURQUOISE
Zuchtform / Breeding form

Photo: H.J. Mayland

S73180-4 *Symphysodon aequifasciatus* Pellegrin, 1904
DISCUS TÜRKIS/TURQUOISE
Zuchtform / Breeding form

Photo: H.J. Mayland

S73181-4 *Symphysodon aequifasciatus* Pellegrin, 1904
DISCUS TÜRKIS/TURQUOISE
Zuchtform / Breeding form

Photo: H.J. Mayland

S73182-4 *Symphysodon aequifasciatus* Pellegrin, 1904
DISCUS TÜRKIS/TURQUOISE
Zuchtform / Breeding form

Photo: H.J. Mayland

S73183-4 *Symphysodon aequifasciatus* Pellegrin, 1904
DISCUS TÜRKIS/TURQUOISE
Zuchtform / Breeding form

Photo: Archiv A.C.S.

S73184-4 *Symphysodon aequifasciatus* Pellegrin, 1904
DISCUS TÜRKIS/TURQUOISE
Zuchtform / Breeding form

Photo: H.J. Mayland

Gruppe 5 / Group 5: Türkis gezeichnete Diskus / Turquoise Striped Discus

Gruppe 5 / Group 5: Türkis gezeichnete Diskus / Turquoise Striped Discus

S73185-4 *Symphysodon aequifasciatus* PELLEGRIN, 1904
DISCUS TÜRKIS/TURQUOISE
Zuchtform / Breeding form

Photo: J. Stendker

S73186-4 *Symphysodon aequifasciatus* PELLEGRIN, 1904
DISCUS TÜRKIS/TURQUOISE
Zuchtform / Breeding form

Photo: J. Stendker

S73187-4 *Symphysodon aequifasciatus* PELLEGRIN, 1904
DISCUS TÜRKIS/TURQUOISE
Zuchtform / Breeding form

Photo: Archiv A.C.S.

S73188-4 *Symphysodon aequifasciatus* PELLEGRIN, 1904
DISCUS TÜRKIS/TURQUOISE
Zuchtform / Breeding form

Photo: Archiv A.C.S.

S73189-4 *Symphysodon aequifasciatus* PELLEGRIN, 1904
DISCUS TÜRKIS/TURQUOISE
Zuchtform / Breeding form

Photo: Archiv A.C.S.

S73190-4 *Symphysodon aequifasciatus* PELLEGRIN, 1904
DISCUS TÜRKIS/TURQUOISE
Zuchtform / Breeding form

Photo: Takrit-Aquarium Bangkok

S73192-4 *Symphysodon aequifasciatus* PELLEGRIN, 1904
DISCUS TÜRKIS/TURQUOISE
Zuchtform / Breeding form

Photo: U. Werner

S73194-4 *Symphysodon aequifasciatus* PELLEGRIN, 1904
DISCUS TÜRKIS/TURQUOISE
Zuchtform / Breeding form

Photo: Archiv A.C.S.

Gruppe 5 / Group 5: Türkis gezeichnete Diskus / Turquoise Striped Discus

S73195-4 *Symphysodon aequifasciatus* PELLEGRIN, 1904
DISCUS TÜRKIS/TURQUOISE
Zuchtform / Breeding form

Photo: Tomizana / A.C.S.

South American **Cichlids IV** — **123**

Gruppe 5 / Group 5: Türkis gezeichnete Diskus / Turquoise Striped Discus

S73197-4 *Symphysodon aequifasciatus* Pellegrin, 1904
DISCUS TÜRKIS/TURQUOISE
Zuchtform / Breeding form

Photo: M. Göbel

S73198-4 *Symphysodon aequifasciatus* Pellegrin, 1904
DISCUS TÜRKIS/TURQUOISE
Zuchtform / Breeding form

Photo: ACS/TFF Mainland

S73199-4 *Symphysodon aequifasciatus* Pellegrin, 1904
DISCUS TÜRKIS/TURQUOISE
Zuchtform / Breeding form

Photo: M. Göbel

S73200-4 *Symphysodon aequifasciatus* Pellegrin, 1904
DISCUS TÜRKIS/TURQUOISE
Zuchtform / Breeding form

Photo: H. Nakano

S73201-4 *Symphysodon aequifasciatus* Pellegrin, 1904
DISCUS TÜRKIS/TURQUOISE
Zuchtform / Breeding form

Photo: M. Göbel

S73202-4 *Symphysodon aequifasciatus* Pellegrin, 1904
DISCUS TÜRKIS/TURQUOISE
Zuchtform / Breeding form

Photo: M. Göbel

S73203-4 *Symphysodon aequifasciatus* PELLEGRIN, 1904
DISCUS TÜRKIS/TURQUOISE
Zuchtform / Breeding form

Photo: J. Glaser

S73204-4 *Symphysodon aequifasciatus* PELLEGRIN, 1904
DISCUS TÜRKIS/TURQUOISE
Zuchtform / Breeding form

Photo: H.J. Mayland

S73205-4 *Symphysodon aequifasciatus* PELLEGRIN, 1904
DISCUS TÜRKIS/TURQUOISE
Zuchtform / Breeding form

Photo: H.J. Mayland

S73206-4 *Symphysodon aequifasciatus* PELLEGRIN, 1904
DISCUS TÜRKIS/TURQUOISE
Zuchtform / Breeding form

Photo: H.J. Mayland

S73207-4 *Symphysodon aequifasciatus* PELLEGRIN, 1904
DISCUS TÜRKIS/TURQUOISE mit Gelege / with eggs
Zuchtform / Breeding form

Photo: H.J. Mayland

S73208-4 *Symphysodon aequifasciatus* PELLEGRIN, 1904
DISCUS TÜRKIS/TURQUOISE
Zuchtform / Breeding form

Photo: Archiv A.C.S.

Gruppe 5 / Group 5: Türkis gezeichnete Diskus / Turquoise Striped Discus

South American Cichlids IV

Gruppe 5 / Group 5: Türkis gezeichnete Diskus / Turquoise Striped Discus

S73209-4 *Symphysodon aequifasciatus* PELLEGRIN, 1904
DISCUS TÜRKIS/TURQUOISE
Zuchtform / Breeding form

Photo: H.J. Mayland

S73210-4 *Symphysodon aequifasciatus* PELLEGRIN, 1904
DISCUS TÜRKIS/TURQUOISE
Zuchtform / Breeding form

Photo: H.J. Mayland

S73211-4 *Symphysodon aequifasciatus* PELLEGRIN, 1904
DISCUS TÜRKIS/TURQUOISE
Zuchtform / Breeding form

Photo: W.A. Tomey

S73212-4 *Symphysodon aequifasciatus* PELLEGRIN, 1904
DISCUS TÜRKIS/TURQUOISE
Zuchtform / Breeding form

Photo: M. Göbel

S73213-4 *Symphysodon aequifasciatus* PELLEGRIN, 1904
DISCUS TÜRKIS/TURQUOISE
Zuchtform / Breeding form

Photo: M. Göbel

S73214-4 *Symphysodon aequifasciatus* PELLEGRIN, 1904
DISCUS TÜRKIS/TURQUOISE
Zuchtform / Breeding form

Photo: M. Göbel

S90501-4 *Symphysodon aequifasciatus* Pellegrin, 1904
DISCUS GREEN-BLUE flächig/solid
Zuchtform / Breeding form

Photo: J. Stendker

S90502-4 *Symphysodon aequifasciatus* Pellegrin, 1904
DISCUS GREEN-BLUE flächig/solid
Zuchtform / Breeding form

Photo: J. Stendker

S90503-4 *Symphysodon aequifasciatus* Pellegrin, 1904
DISCUS GREEN-BLUE flächig/solid
Zuchtform / Breeding form

Photo: J. Stendker

S90504-4 *Symphysodon aequifasciatus* Pellegrin, 1904
DISCUS GREEN-BLUE flächig/solid
Zuchtform / Breeding form

Photo: J. Stendker

S90505-4 *Symphysodon aequifasciatus* Pellegrin, 1904
DISCUS GREEN-BLUE flächig/solid
Zuchtform / Breeding form

Photo: J. Stendker

Gruppe 5 / Group 5: Türkis gezeichnete Diskus / Turquoise Striped Discus

Gruppe 6 / Group 6: Discus flächig grün/blau/türkis / Discus solid green/blue/turquoise

S90506-4 *Symphysodon aequifasciatus* Pellegrin, 1904
DISCUS - SOLID TURQUOISE
Zuchtform / Breeding form -1971-

Photo: M. Göbel

S90507-4 *Symphysodon aequifasciatus* Pellegrin, 1904
DISCUS - SOLID TURQUOISE
Zuchtform / Breeding form -1971-

Photo: M. Göbel

S90508-4 *Symphysodon aequifasciatus* Pellegrin, 1904
DISCUS - SOLID TURQUOISE
Zuchtform / Breeding form

Photo: M. Göbel

S90532-4 *Symphysodon aequifasciatus* Pellegrin, 1904
DISCUS - SOLID TURQUOISE
Zuchtform / Breeding form

Photo: M. Göbel

S90547-4 *Symphysodon aequifasciatus* Pellegrin, 1904
DISCUS - SOLID TURQUOISE
Zuchtform / Breeding form

Photo: H.J. Mayland

S90549-4 *Symphysodon aequifasciatus* Pellegrin, 1904
DISCUS - SOLID TURQUOISE
Zuchtform / Breeding form

Photo: M. Göbel

S90509-4 *Symphysodon aequifasciatus* Pellegrin, 1904
DISCUS flächig blau/türkis / solid blue/turquoise
Zuchtform / Breeding form

Photo: H.J. Mayland

S90510-4 *Symphysodon aequifasciatus* Pellegrin, 1904
DISCUS flächig blau/türkis / solid blue/turquoise
Zuchtform / Breeding form

Photo: M. Göbel

S90511-4 *Symphysodon aequifasciatus* Pellegrin, 1904
DISCUS flächig blau/türkis / solid blue/turquoise
Zuchtform / Breeding form

Photo: H. Nakano

S90512-4 *Symphysodon aequifasciatus* Pellegrin, 1904
DISCUS flächig blau/türkis / solid blue/turquoise
Zuchtform / Breeding form

Photo: H. Nakano

S90513-4 *Symphysodon aequifasciatus* Pellegrin, 1904
DISCUS flächig blau/türkis / solid blue/turquoise
Zuchtform / Breeding form

Photo: Schlingmann

S90514-4 *Symphysodon aequifasciatus* Pellegrin, 1904
DISCUS flächig blau/türkis / solid blue/turquoise
Zuchtform / Breeding form

Photo: M. Göbel

Gruppe 6 / Group 6: Discus flächig grün/blau/türkis / Discus solid green/blue/turquoise

Gruppe 6 / Group 6: Discus flächig grün/blau/türkis / Discus solid green/blue/turquoise

S90515-4 *Symphysodon aequifasciatus* Pellegrin, 1904
DISCUS flächig blau/türkis / solid blue/turquoise
Zuchtform / Breeding form

Photo: H.J. Mayland

S90516-4 *Symphysodon aequifasciatus* Pellegrin, 1904
DISCUS flächig blau/türkis / solid blue/turquoise
Zuchtform / Breeding form

Photo: U. Werner

S90517-4 *Symphysodon aequifasciatus* Pellegrin, 1904
DISCUS flächig blau/türkis / solid blue/turquoise
Zuchtform / Breeding form

Photo: Cheun Wai Shing

S90518-4 *Symphysodon aequifasciatus* Pellegrin, 1904
DISCUS flächig blau/türkis / solid blue/turquoise
Zuchtform / Breeding form

Photo: Cheun Wai Shing

S90519-4 *Symphysodon aequifasciatus* Pellegrin, 1904
DISCUS flächig blau/türkis / solid blue/turquoise
Zuchtform / Breeding form

Photo: H.J. Mayland

S90520-4 *Symphysodon aequifasciatus* Pellegrin, 1904
DISCUS flächig blau/türkis / solid blue/turquoise
Zuchtform / Breeding form

Photo: M. Göbel

S90524-4 *Symphysodon aequifasciatus* Pellegrin, 1904
DISCUS flächig blau/türkis / solid blue/turquoise
Zuchtform / Breeding form

Photo: Takrit-Aquarium, Bangkok

S90522-4 *Symphysodon aequifasciatus* Pellegrin, 1904
DISCUS flächig blau/türkis / solid blue/turquoise
Zuchtform / Breeding form

Photo: Archiv A.C.S.

Gruppe 6 / Group 6: Discus flächig grün/blau/türkis / Discus solid green/blue/turquoise

S90525-4 *Symphysodon aequifasciatus* Pellegrin, 1904
DISCUS flächig blau/türkis / solid blue/turquoise
"REFLECTION-DISCUS", Zuchtform / Breeding form

Photo: Wayne DC HongKong

South American Cichlids IV — 131

Gruppe 6 / Group 6: Discus flächig grün/blau/türkis / Discus solid green/blue/turquoise

S90527-4 *Symphysodon aequifasciatus* Pellegrin, 1904
DISCUS flächig blau/türkis / solid blue/turquoise
Zuchtform / Breeding form

Photo: J. Stendker

S90528-4 *Symphysodon aequifasciatus* Pellegrin, 1904
DISCUS flächig blau/türkis / solid blue/turquoise
Zuchtform / Breeding form

Photo: J. Stendker

S90529-4 *Symphysodon aequifasciatus* Pellegrin, 1904
DISCUS flächig blau/türkis / solid blue/turquoise
Zuchtform / Breeding form

Photo: Archiv A.C.S.

S90530-4 *Symphysodon aequifasciatus* Pellegrin, 1904
DISCUS flächig blau/türkis / solid blue/turquoise
Zuchtform / Breeding form

Photo: J. Stendker

S90531-4 *Symphysodon aequifasciatus* Pellegrin, 1904
DISCUS flächig blau/türkis / solid blue/turquoise
Zuchtform / Breeding form

Photo: J. Stendker

S90531-4 *Symphysodon aequifasciatus* Pellegrin, 1904
DISCUS flächig blau/türkis / solid blue/turquoise
Zuchtform / Breeding form

Photo: J. Stendker

S90533-4 *Symphysodon aequifasciatus* Pellegrin, 1904
DISCUS flächig blau/türkis / solid blue/turquoise
Zuchtform / Breeding form

Photo: M. Göbel

S90534-4 *Symphysodon aequifasciatus* Pellegrin, 1904
DISCUS flächig blau/türkis / solid blue/turquoise
Zuchtform / Breeding form

Photo: M. Göbel

S90535-4 *Symphysodon aequifasciatus* Pellegrin, 1904
DISCUS flächig blau/türkis / solid blue/turquoise
Zuchtform / Breeding form

Photo: M. Göbel

S90536-4 *Symphysodon aequifasciatus* Pellegrin, 1904
DISCUS flächig blau/türkis / solid blue/turquoise
Zuchtform / Breeding form

Photo: M. Göbel

S90537-4 *Symphysodon aequifasciatus* Pellegrin, 1904
DISCUS flächig blau/türkis / solid blue/turquoise
Zuchtform / Breeding form

Photo: M. Göbel

S90538-4 *Symphysodon aequifasciatus* Pellegrin, 1904
DISCUS flächig blau/türkis / solid blue/turquoise
Zuchtform / Breeding form

Photo: M. Göbel

Gruppe 6 / Group 6: Discus flächig grün/blau/türkis / Discus solid green/blue/turquoise

Gruppe 6 / Group 6: Discus flächig grün/blau/türkis / Discus solid green/blue/turquoise

S90539-4 *Symphysodon aequifasciatus* PELLEGRIN, 1904
DISCUS flächig blau/türkis / solid blue/turquoise
"BLUE-DIAMOND", Zuchtform / Breeding form
Photo: Nakano / Archiv A.C.S.

S90540-4 *Symphysodon aequifasciatus* PELLEGRIN, 1904
DISCUS flächig blau/türkis / solid blue/turquoise
"BLUE-DIAMOND", Zuchtform / Breeding form
Photo: H.J. Mayland

S90541-4 *Symphysodon aequifasciatus* PELLEGRIN, 1904
DISCUS flächig blau/türkis / solid blue/turquoise
"BLUE-DIAMOND", Zuchtform / Breeding form
Photo: H.J. Mayland

S90542-4 *Symphysodon aequifasciatus* PELLEGRIN, 1904
DISCUS flächig blau/türkis / solid blue/turquoise
"BLUE-DIAMOND", Zuchtform / Breeding form
Photo: ACS/TFF Mainland

S90543-4 *Symphysodon aequifasciatus* PELLEGRIN, 1904
DISCUS flächig blau/türkis / solid blue/turquoise
"BLUE-DIAMOND", Zuchtform / Breeding form
Photo: Wayne DC HongKong

S90544-4 *Symphysodon aequifasciatus* PELLEGRIN, 1904
DISCUS flächig blau/türkis / solid blue/turquoise
"BLUE-DIAMOND", Zuchtform / Breeding form
Photo: H.J. Mayland

S90538-4 *Symphysodon aequifasciatus* Pellegrin, 1904
DISCUS flächig blau/türkis / solid blue/turquoise
Zuchtform / Breeding form

Gruppe 6 / Group 6: Discus flächig grün/blau/türkis / Discus solid green/blue/turquoise

S90545-4 *Symphysodon aequifasciatus* Pellegrin, 1904
DISCUS flächig blau/türkis / solid blue/turquoise
"BLUE-DIAMOND", Zuchtform / Breeding form

Photo: Wayne DC HongKong

S90546-4 *Symphysodon aequifasciatus* Pellegrin, 1904
DISCUS flächig blau/türkis / solid blue/turquoise
"BLUE-DIAMOND", Zuchtform / Breeding form

Photo: Cheun Wai Shing

S90548-4 *Symphysodon aequifasciatus* Pellegrin, 1904
DISCUS flächig blau/türkis / solid blue/turquoise "BLUE DIAMOND"
Zuchtform / Breeding form

Sieger/best of Show
-AQUARAMA Singapore 1997-

Photo: Cheun Wai Shing

S74001-4 *Symphysodon aequifasciatus* Pellegrin, 1904
DISCUS ROT-TÜRKIS / RED TURQUOISE
Zuchtform / Breeding form

Photo: H.J. Mayland

S74002-4 *Symphysodon aequifasciatus* Pellegrin, 1904
DISCUS ROT-TÜRKIS / RED TURQUOISE
Zuchtform / Breeding form

Photo: H.J. Mayland

S74003-4 *Symphysodon aequifasciatus* Pellegrin, 1904
DISCUS ROT-TÜRKIS / RED TURQUOISE
Zuchtform / Breeding form

Photo: H.J. Mayland

S74004-4 *Symphysodon aequifasciatus* Pellegrin, 1904
DISCUS ROT-TÜRKIS / RED TURQUOISE
Zuchtform / Breeding form

Photo: H.J. Mayland

S74005-4 *Symphysodon aequifasciatus* Pellegrin, 1904
DISCUS ROT-TÜRKIS / RED TURQUOISE
Zuchtform / Breeding form

Photo: H.J. Mayland

S74006-4 *Symphysodon aequifasciatus* Pellegrin, 1904
DISCUS ROT-TÜRKIS / RED TURQUOISE
Zuchtform / Breeding form

Photo: H.J. Mayland

Gruppe 7 / Group 7: Rot-türkis / Red-turquoise

Gruppe 7 / Group 7: Rot-türkis / Red-turquoise

S74007-4 *Symphysodon aequifasciatus* PELLEGRIN, 1904
DISCUS ROT-TÜRKIS / RED TURQUOISE
Zuchtform / Breeding form

Photo: U. Werner

S74008-4 *Symphysodon aequifasciatus* PELLEGRIN, 1904
DISCUS ROT-TÜRKIS / RED TURQUOISE
Zuchtform / Breeding form

Photo: H.J. Mayland

S74009-4 *Symphysodon aequifasciatus* PELLEGRIN, 1904
DISCUS ROT-TÜRKIS / RED TURQUOISE
Zuchtform / Breeding form

Photo: W. Mikschofsky

S74010-4 *Symphysodon aequifasciatus* PELLEGRIN, 1904
DISCUS ROT-TÜRKIS / RED TURQUOISE
Zuchtform / Breeding form

Photo: W. Mikschofsky

S74011-4 *Symphysodon aequifasciatus* PELLEGRIN, 1904
DISCUS ROT-TÜRKIS / RED TURQUOISE
Zuchtform / Breeding form

Photo: ACS/TFF Mainland

S74012-4 *Symphysodon aequifasciatus* PELLEGRIN, 1904
DISCUS ROT-TÜRKIS / RED TURQUOISE
Zuchtform / Breeding form

Photo: ACS/TFF Mainland

S74013-4 *Symphysodon aequifasciatus* Pellegrin, 1904
DISCUS ROT-TÜRKIS / RED TURQUOISE
Zuchtform / Breeding form

Photo: H.J. Mayland

S74014-4 *Symphysodon aequifasciatus* Pellegrin, 1904
DISCUS ROT-TÜRKIS / RED TURQUOISE
Zuchtform / Breeding form

Photo: H.J. Mayland

S74015-4 *Symphysodon aequifasciatus* Pellegrin, 1904
DISCUS ROT-TÜRKIS / RED TURQUOISE
Zuchtform / Breeding form

Photo: M. Göbel

S74016-4 *Symphysodon aequifasciatus* Pellegrin, 1904
DISCUS ROT-TÜRKIS / RED TURQUOISE
Zuchtform / Breeding form

Photo: M. Göbel

S74017-4 *Symphysodon aequifasciatus* Pellegrin, 1904
DISCUS ROT-TÜRKIS / RED TURQUOISE
Zuchtform / Breeding form

Photo: M. Göbel

S74018-4 *Symphysodon aequifasciatus* Pellegrin, 1904
DISCUS ROT-TÜRKIS / RED TURQUOISE
Zuchtform / Breeding form

Photo: M. Göbel

Gruppe 7 / Group 7: Rot-türkis / Red -turquoise

Gruppe 7 / Group 7: Rot-türkis / Red-turquoise

S74019-4 *Symphysodon aequifasciatus* Pellegrin, 1904
DISCUS ROT-TÜRKIS / RED TURQUOISE
Zuchtform / Breeding form

Photo: W. Mikschofsky

S74020-4 *Symphysodon aequifasciatus* Pellegrin, 1904
DISCUS ROT-TÜRKIS / RED TURQUOISE
Zuchtform / Breeding form

Photo: W. Mikschofsky

S74021-4 *Symphysodon aequifasciatus* Pellegrin, 1904
DISCUS ROT-TÜRKIS / RED TURQUOISE
Zuchtform / Breeding form

Photo: W. Mikschofsky

S74022-4 *Symphysodon aequifasciatus* Pellegrin, 1904
DISCUS ROT-TÜRKIS / RED TURQUOISE
Zuchtform / Breeding form

Photo: W. Mikschofsky

S74023-4 *Symphysodon aequifasciatus* Pellegrin, 1904
DISCUS ROT-TÜRKIS / RED TURQUOISE
Zuchtform / Breeding form

Photo: W. Mikschofsky

S74024-4 *Symphysodon aequifasciatus* Pellegrin, 1904
DISCUS ROT-TÜRKIS / RED TURQUOISE
Zuchtform / Breeding form

Photo: W. Mikschofsky

S74082-4 *Symphysodon aequifasciatus* Pellegrin, 1904
DISCUS ROT-TÜRKIS / RED TURQUOISE
Zuchtform / Breeding form

Photo: F. Schulten

S74083-3 *Symphysodon aequifasciatus* Pellegrin, 1904
DISCUS ROT-TÜRKIS / RED TURQUOISE
Brazil, Zuchtform/breed. form -semiadult-

Photo: H. Nakano

S74084-4 *Symphysodon aequifasciatus* Pellegrin, 1904
DISCUS ROT-TÜRKIS / RED TURQUOISE
Zuchtform / Breeding form

Photo: H. Nakano

S74085-4 *Symphysodon aequifasciatus* Pellegrin, 1904
DISCUS ROT-TÜRKIS / RED TURQUOISE
Zuchtform / Breeding form

Photo: F. Schulten

S74086-4 *Symphysodon aequifasciatus* Pellegrin, 1904
DISCUS ROT-TÜRKIS / RED TURQUOISE
Zuchtform / Breeding form

Photo: F. Schulten

S74087-4 *Symphysodon aequifasciatus* Pellegrin, 1904
DISCUS ROT-TÜRKIS / RED TURQUOISE
Zuchtform / Breeding form

Photo: F. Schulten

Gruppe 7 / Group 7: Rot-türkis / Red -turquoise

Gruppe 7 / Group 7: Rot-türkis / Red-turquoise

S74025-4 *Symphysodon aequifasciatus* Pellegrin, 1904
DISCUS ROT-TÜRKIS / RED TURQUOISE
Zuchtform/breeding form

Photo: F. Bodenmüller

S74025-4 *Symphysodon aequifasciatus* Pellegrin, 1904
DISCUS ROT-TÜRKIS / RED TURQUOISE
Zuchtform/breeding form

Photo: F. Bodenmüller

S74027-4 *Symphysodon aequifasciatus* Pellegrin, 1904
DISCUS ROT-TÜRKIS / RED TURQUOISE
Zuchtform/breeding form

Photo: F. Bodenmüller

S74028-4 *Symphysodon aequifasciatus* Pellegrin, 1904
DISCUS ROT-TÜRKIS / RED TURQUOISE
Zuchtform/breeding form

Photo: F. Bodenmüller

S74029-4 *Symphysodon aequifasciatus* Pellegrin, 1904
DISCUS ROT-TÜRKIS / RED TURQUOISE
Zuchtform/breeding form (aus Royal-Blue)

Photo: F. Bodenmüller

S74030-4 *Symphysodon aequifasciatus* Pellegrin, 1904
DISCUS ROT-TÜRKIS / RED TURQUOISE
Zuchtform/breeding form (aus Royal-Blue)

Photo: F. Bodenmüller

S74031-4 *Symphysodon aequifasciatus* Pellegrin, 1904
DISCUS ROT-TÜRKIS / RED TURQUOISE
Zuchtform/breeding form

Photo: M. Göbel

S74032-4 *Symphysodon aequifasciatus* Pellegrin, 1904
DISCUS ROT-TÜRKIS / RED TURQUOISE
Zuchtform/breeding form

Photo: M. Göbel

S74033-4 *Symphysodon aequifasciatus* Pellegrin, 1904
DISCUS ROT-TÜRKIS / RED TURQUOISE
Zuchtform/breeding form

Photo: M. Göbel

S74034-4 *Symphysodon aequifasciatus* Pellegrin, 1904
DISCUS ROT-TÜRKIS / RED TURQUOISE
Zuchtform/breeding form

Photo: M. Göbel

S74035-4 *Symphysodon aequifasciatus* Pellegrin, 1904
DISCUS ROT-TÜRKIS / RED TURQUOISE
Zuchtform/breeding form

Photo: M. Göbel

S74036-4 *Symphysodon aequifasciatus* Pellegrin, 1904
DISCUS ROT-TÜRKIS / RED TURQUOISE
Zuchtform/breeding form

Photo: M. Göbel

Gruppe 7 / Group 7: Rot-türkis / Red-turquoise

S74030-4 *Symphysodon aequifasciatus* Pellegrin, 1904
DISCUS ROT-TÜRKIS / RED TURQUOISE
Zuchtform/breeding form (aus Royal-Blue)

Photo: F. Bodenmüller

S74033-4 *Symphysodon aequifasciatus* Pellegrin, 1904
DISCUS ROT-TÜRKIS / RED-TURQUOISE
Zuchtform/breeding form

Photo: M. Göbel

Gruppe 7 / Group 7: Rot-türkis / Red -turquoise

S74037-4 *Symphysodon aequifasciatus* Pellegrin, 1904
DISCUS ROT-TÜRKIS / RED -TURQUOISE
Zuchtform/breeding form

Photo: J. Stendker

S74038-4 *Symphysodon aequifasciatus* Pellegrin, 1904
DISCUS ROT-TÜRKIS / RED -TURQUOISE
Zuchtform/breeding form

Photo: J. Stendker

S74039-4 *Symphysodon aequifasciatus* Pellegrin, 1904
DISCUS ROT-TÜRKIS / RED -TURQUOISE
Zuchtform/breeding form

Photo: J. Stendker

S74040-4 *Symphysodon aequifasciatus* Pellegrin, 1904
DISCUS ROT-TÜRKIS / RED -TURQUOISE
Zuchtform/breeding form

Photo: J. Stendker

S74041-4 *Symphysodon aequifasciatus* Pellegrin, 1904
DISCUS ROT-TÜRKIS / RED -TURQUOISE
Zuchtform/breeding form

Photo: J. Stendker

S74042-4 *Symphysodon aequifasciatus* Pellegrin, 1904
DISCUS ROT-TÜRKIS / RED -TURQUOISE
Zuchtform/breeding form

Photo: J. Stendker

South American **Cichlids** *IV*   © Verlag A.C.S. GmbH

S74043-4 *Symphysodon aequifasciatus* Pellegrin, 1904
DISCUS ROT-TÜRKIS / RED -TURQUOISE
Zuchtform/breeding form

Photo: J. Stendker

S74044-4 *Symphysodon aequifasciatus* Pellegrin, 1904
DISCUS ROT-TÜRKIS / RED -TURQUOISE
Zuchtform/breeding form

Photo: J. Stendker

S74045-4 *Symphysodon aequifasciatus* Pellegrin, 1904
DISCUS ROT-TÜRKIS / RED -TURQUOISE
Zuchtform/breeding form

Photo: J. Stendker

S74046-4 *Symphysodon aequifasciatus* Pellegrin, 1904
DISCUS ROT-TÜRKIS / RED -TURQUOISE
Zuchtform/breeding form

Photo: J. Stendker

S74047-4 *Symphysodon aequifasciatus* Pellegrin, 1904
DISCUS ROT-TÜRKIS / RED -TURQUOISE Pair
Zuchtform/breeding form (aus Royal-Blue F 1)

Photo: F. Bodenmüller

S74048-4 *Symphysodon aequifasciatus* Pellegrin, 1904
DISCUS ROT-TÜRKIS / RED -TURQUOISE Pair
Zuchtform/breeding form

Photo: F. Bodenmüller

Gruppe 7 / Group 7: Rot-türkis / Red -turquoise

Gruppe 7 / Group 7: Rot-türkis / Red-turquoise

S74049-4 *Symphysodon aequifasciatus* PELLEGRIN, 1904
DISCUS ROT-TÜRKIS / RED -TURQUOISE
Zuchtform/breeding form

Photo: Archiv A.C.S.

S74050-4 *Symphysodon aequifasciatus* PELLEGRIN, 1904
DISCUS ROT-TÜRKIS / RED -TURQUOISE
Zuchtform/breeding form

Photo: Archiv A.C.S.

S74051-4 *Symphysodon aequifasciatus* PELLEGRIN, 1904
DISCUS ROT-TÜRKIS / RED -TURQUOISE
Zuchtform/breeding form

Photo: Archiv A.C.S.

S74052-4 *Symphysodon aequifasciatus* PELLEGRIN, 1904
DISCUS ROT-TÜRKIS / RED -TURQUOISE
Zuchtform/breeding form

Photo: Archiv A.C.S.

S74053-4 *Symphysodon aequifasciatus* PELLEGRIN, 1904
DISCUS ROT-TÜRKIS / RED -TURQUOISE
Zuchtform/breeding form

Photo: Archiv A.C.S.

S74054-4 *Symphysodon aequifasciatus* PELLEGRIN, 1904
DISCUS ROT-TÜRKIS / RED -TURQUOISE
Zuchtform/breeding form

Photo: Archiv A.C.S.

S74055-4 *Symphysodon aequifasciatus* Pellegrin, 1904
DISCUS ROT-TÜRKIS / RED -TURQUOISE
Zuchtform/breeding form

Photo: W. Mikschofsky

S74056-4 *Symphysodon aequifasciatus* Pellegrin, 1904
DISCUS ROT-TÜRKIS / RED -TURQUOISE
Zuchtform/breeding form

Photo: W. Mikschofsky

S74057-4 *Symphysodon aequifasciatus* Pellegrin, 1904
DISCUS ROT-TÜRKIS / RED -TURQUOISE
Zuchtform/breeding form

Photo: Archiv A.C.S.

S74058-4 *Symphysodon aequifasciatus* Pellegrin, 1904
DISCUS ROT-TÜRKIS / RED -TURQUOISE
Zuchtform/breeding form

Photo: Archiv A.C.S.

S74059-4 *Symphysodon aequifasciatus* Pellegrin, 1904
DISCUS ROT-TÜRKIS / RED -TURQUOISE
Zuchtform/breeding form

Photo: Archiv A.C.S.

S74060-4 *Symphysodon aequifasciatus* Pellegrin, 1904
DISCUS ROT-TÜRKIS / RED -TURQUOISE
Zuchtform/breeding form

Photo: Archiv A.C.S.

Gruppe 7 / Group 7: Rot-türkis / Red -turquoise

Gruppe 7 / Group 7: Rot-türkis / Red-turquoise

S74061-4 *Symphysodon aequifasciatus* Pellegrin, 1904
DISCUS ROT-TÜRKIS / RED -TURQUOISE
Zuchtform/breeding form

Photo: Dr. Lim Yu Hoe

S74062-4 *Symphysodon aequifasciatus* Pellegrin, 1904
DISCUS ROT-TÜRKIS / RED -TURQUOISE
Zuchtform/breeding form

Photo: Dr. Lim Yu Hoe

S74063-4 *Symphysodon aequifasciatus* Pellegrin, 1904
DISCUS ROT-TÜRKIS / RED -TURQUOISE
Zuchtform/breeding form

Photo: Schlingmann

S74064-4 *Symphysodon aequifasciatus* Pellegrin, 1904
DISCUS ROT-TÜRKIS / RED -TURQUOISE
Zuchtform/breeding form

Photo: Archiv A.C.S.

S74065-4 *Symphysodon aequifasciatus* Pellegrin, 1904
DISCUS ROT-TÜRKIS / RED -TURQUOISE
Zuchtform/breeding form

Photo: Archiv A.C.S.

S74066-4 *Symphysodon aequifasciatus* Pellegrin, 1904
DISCUS ROT-TÜRKIS / RED -TURQUOISE
Zuchtform/breeding form

Photo: Archiv A.C.S.

S74067-4 *Symphysodon aequifasciatus* PELLEGRIN, 1904
DISCUS ROT-TÜRKIS / RED -TURQUOISE
Zuchtform/breeding form

Photo: Archiv A.C.S.

S74068-4 *Symphysodon aequifasciatus* PELLEGRIN, 1904
DISCUS ROT-TÜRKIS / RED -TURQUOISE
Zuchtform/breeding form

Photo: M. Göbel

S74069-4 *Symphysodon aequifasciatus* PELLEGRIN, 1904
DISCUS ROT-TÜRKIS / RED -TURQUOISE
Zuchtform/breeding form

Photo: M. Göbel

S74070-4 *Symphysodon aequifasciatus* PELLEGRIN, 1904
DISCUS ROT-TÜRKIS / RED -TURQUOISE
Zuchtform/breeding form

Photo: M. Göbel

S74071-4 *Symphysodon aequifasciatus* PELLEGRIN, 1904
DISCUS ROT-TÜRKIS / RED -TURQUOISE
Zuchtform/breeding form

Photo: M. Göbel

S74072-4 *Symphysodon aequifasciatus* PELLEGRIN, 1904
DISCUS ROT-TÜRKIS / RED -TURQUOISE
Zuchtform/breeding form

Photo: M. Göbel

Gruppe 7 / Group 7: Rot-türkis / Red -turquoise

Gruppe 7 / Group 7: Rot-türkis / Red-turquoise

S74073-4 *Symphysodon aequifasciatus* Pellegrin, 1904
DISCUS ROT-TÜRKIS / RED -TURQUOISE
Zuchtform/breeding form

Photo: Wayne DC Hkg.

S74074-4 *Symphysodon aequifasciatus* Pellegrin, 1904
DISCUS ROT-TÜRKIS / RED -TURQUOISE
Zuchtform/breeding form

Photo: Cheun Wai Shing

S74075-4 *Symphysodon aequifasciatus* Pellegrin, 1904
DISCUS ROT-TÜRKIS / RED -TURQUOISE
Zuchtform/breeding form

Photo: Cheun Wai Shing

S74076-4 *Symphysodon aequifasciatus* Pellegrin, 1904
DISCUS ROT-TÜRKIS / RED -TURQUOISE
Zuchtform/breeding form

Photo: Cheun Wai Shing

S74077-4 *Symphysodon aequifasciatus* Pellegrin, 1904
DISCUS ROT-TÜRKIS / RED -TURQUOISE
Zuchtform/breeding form

Photo: Cheun Wai Shing

S74078-4 *Symphysodon aequifasciatus* Pellegrin, 1904
DISCUS ROT-TÜRKIS / RED -TURQUOISE
Zuchtform/breeding form

Photo: Cheun Wai Shing

S74079-4 *Symphysodon aequifasciatus* Pellegrin, 1904
DISCUS ROT-TÜRKIS / RED -TURQUOISE
Zuchtform/breeding form

Photo: Archiv A.C.S.

S74120-4 *Symphysodon aequifasciatus* Pellegrin, 1904
DISCUS ROT-TÜRKIS / RED -TURQUOISE
Zuchtform/breeding form

Photo: Schlingmann

S74121-4 *Symphysodon aequifasciatus* Pellegrin, 1904
DISCUS ROT-TÜRKIS / RED -TURQUOISE
Zuchtform/breeding form

Photo: Dr. Lim Yu Hoe

S74122-4 *Symphysodon aequifasciatus* Pellegrin, 1904
DISCUS ROT-TÜRKIS / RED -TURQUOISE
"Best of Show (Grand Champion) Aquarama 1993"

Photo: Dr. Lim Yu Hoe

S74123-4 *Symphysodon aequifasciatus* Pellegrin, 1904
DISCUS ROT-TÜRKIS / RED -TURQUOISE
Zuchtform/breeding form

Photo: Dr. Lim Yu Hoe

S74124-4 *Symphysodon aequifasciatus* Pellegrin, 1904
DISCUS ROT-TÜRKIS / RED -TURQUOISE
Zuchtform/breeding form

Photo: Dr. Lim Yu Hoe

Gruppe 7 / Group 7: Rot-türkis / Red -turquoise

Gruppe 7 / Group 7: Rot-türkis / Red-turquoise

S74111-4 *Symphysodon aequifasciatus* PELLEGRIN, 1904
DISCUS ROT-TÜRKIS / RED -TURQUOISE
Zuchtform/breeding form
Photo: M. Göbel

S74112-4 *Symphysodon aequifasciatus* PELLEGRIN, 1904
DISCUS ROT-TÜRKIS / RED -TURQUOISE
Zuchtform/breeding form
Photo: M. Göbel

S74113-4 *Symphysodon aequifasciatus* PELLEGRIN, 1904
DISCUS ROT-TÜRKIS / RED -TURQUOISE
Zuchtform/breeding form
Photo: M. Göbel

S74114-4 *Symphysodon aequifasciatus* PELLEGRIN, 1904
DISCUS ROT-TÜRKIS / RED -TURQUOISE
Zuchtform/breeding form
Photo: M. Göbel

S74115-4 *Symphysodon aequifasciatus* PELLEGRIN, 1904
DISCUS ROT-TÜRKIS / RED -TURQUOISE
Zuchtform/breeding form
Photo: M. Göbel

S74116-4 *Symphysodon aequifasciatus* PELLEGRIN, 1904
DISCUS ROT-TÜRKIS / RED -TURQUOISE
Zuchtform/breeding form
Photo: M. Göbel

S74088-4 *Symphysodon aequifasciatus* Pellegrin, 1904
DISCUS ROT-TÜRKIS / RED -TURQUOISE
Zuchtform/breeding form

Photo: Archiv A.C.S.

S74089-4 *Symphysodon aequifasciatus* Pellegrin, 1904
DISCUS ROT-TÜRKIS / RED -TURQUOISE
Zuchtform/breeding form

Photo: M. Göbel

S74090-4 *Symphysodon aequifasciatus* Pellegrin, 1904
DISCUS ROT-TÜRKIS / RED -TURQUOISE
Zuchtform/breeding form

Photo: M. Göbel

S74091-4 *Symphysodon aequifasciatus* Pellegrin, 1904
DISCUS ROT-TÜRKIS / RED -TURQUOISE
Zuchtform/breeding form

Photo: M. Göbel

S74092-4 *Symphysodon aequifasciatus* Pellegrin, 1904
DISCUS ROT-TÜRKIS / RED -TURQUOISE
Zuchtform/breeding form

Photo: M. Göbel

S74093-4 *Symphysodon aequifasciatus* Pellegrin, 1904
DISCUS ROT-TÜRKIS / RED -TURQUOISE
Zuchtform/breeding form

Photo: Cheun Wai Shing

Gruppe 7 / Group 7: Rot-türkis / Red -turquoise

Gruppe 7 / Group 7: Rot-türkis / Red-turquoise

S74150-4 *Symphysodon aequifasciatus* PELLEGRIN, 1904
DISCUS ROT-TÜRKIS / RED -TURQUOISE
Zuchtform/breeding form

Photo: M. Göbel

S74151-4 *Symphysodon aequifasciatus* PELLEGRIN, 1904
DISCUS ROT-TÜRKIS / RED -TURQUOISE
Zuchtform/breeding form

Photo: J. Schütz

S74152-4 *Symphysodon aequifasciatus* PELLEGRIN, 1904
DISCUS ROT-TÜRKIS / RED -TURQUOISE
Zuchtform/breeding form

Photo: J. Schütz

S74153-4 *Symphysodon aequifasciatus* PELLEGRIN, 1904
DISCUS ROT-TÜRKIS / RED -TURQUOISE
Zuchtform/breeding form

Photo: J. Schütz

S74154-4 *Symphysodon aequifasciatus* PELLEGRIN, 1904
DISCUS ROT-TÜRKIS / RED -TURQUOISE
Zuchtform/breeding form

Photo: J. Schütz

S74155-4 *Symphysodon aequifasciatus* PELLEGRIN, 1904
DISCUS ROT-TÜRKIS / RED -TURQUOISE
Zuchtform/breeding form

Photo: J. Schütz

S74156-4 *Symphysodon aequifasciatus* Pellegrin, 1904
DISCUS ROT-TÜRKIS / RED-TURQUOISE
Zuchtform/breeding form

Photo: M. Göbel

Gruppe 7 / Group 7: Rot-türkis / Red-turquoise

S74094-4 *Symphysodon aequifasciatus* Pellegrin, 1904
DISCUS ROT-TÜRKIS / RED-TURQUOISE
Zuchtform/breeding form

Photo: M. Göbel

S74095-4 *Symphysodon aequifasciatus* Pellegrin, 1904
DISCUS ROT-TÜRKIS / RED-TURQUOISE
Zuchtform/breeding form

Photo: M. Göbel

S74096-4 *Symphysodon aequifasciatus* Pellegrin, 1904
DISCUS ROT-TÜRKIS / RED-TURQUOISE
Zuchtform/breeding form

Photo: M. Göbel

S74097-4 *Symphysodon aequifasciatus* Pellegrin, 1904
DISCUS ROT-TÜRKIS / RED-TURQUOISE
Zuchtform/breeding form

Photo: M. Göbel

S74098-4 *Symphysodon aequifasciatus* Pellegrin, 1904
DISCUS ROT-TÜRKIS / RED-TURQUOISE
Zuchtform/breeding form

Photo: M. Göbel

S74099-4 *Symphysodon aequifasciatus* Pellegrin, 1904
DISCUS ROT-TÜRKIS / RED-TURQUOISE
Zuchtform/breeding form

Photo: M. Göbel

S74100-4 *Symphysodon aequifasciatus* Pellegrin, 1904
DISCUS ROT-TÜRKIS / RED -TURQUOISE
Zuchtform/breeding form "Tangerine Dream F_1"
Photo: M. Göbel

S74130-4 *Symphysodon aequifasciatus* Pellegrin, 1904
DISCUS ROT-TÜRKIS / RED -TURQUOISE
Zuchtform/breeding form "Tangerine Dream F_1"
Photo: M. Göbel

S74101-4 *Symphysodon aequifasciatus* Pellegrin, 1904
DISCUS ROT-TÜRKIS / RED -TURQUOISE
Zuchtform/breeding form "Tangerine Dream F_2"
Photo: M. Göbel

S74131-4 *Symphysodon aequifasciatus* Pellegrin, 1904
DISCUS ROT-TÜRKIS / RED -TURQUOISE
Zuchtform/breeding form "Tangerine Dream F_3"
Photo: M. Göbel

S74102-4 *Symphysodon aequifasciatus* Pellegrin, 1904
DISCUS ROT-TÜRKIS / RED -TURQUOISE
Zuchtform/breeding form "Tangerine Dream F_3"
Photo: M. Göbel

S74132-4 *Symphysodon aequifasciatus* Pellegrin, 1904
DISCUS ROT-TÜRKIS / RED -TURQUOISE
Zuchtform/breeding form "Tangerine Dream F_3"
Photo: M. Göbel

Gruppe 7 / Group 7: Rot-türkis / Red -turquoise

S74096-4 *Symphysodon aequifasciatus* Pellegrin, 1904
DISCUS ROT-TÜRKIS / RED -TURQUOISE
Zuchtform/breeding form

S74103-4 *Symphysodon aequifasciatus* Pellegrin, 1904
DISCUS ROT-TÜRKIS / RED-TURQUOISE
Zuchtform/breeding form "Red Silk F_1"

♂ ⚠ ℙ ◐ ☺ 🖳 🐟 ⚠ 🔲
Photo: M. Göbel

S74133-4 *Symphysodon aequifasciatus* Pellegrin, 1904
DISCUS ROT-TÜRKIS / RED-TURQUOISE
Zuchtform/breeding form "Red Silk F_1"

♀ ⚠ ℙ ◐ ☺ 🖳 🐟 ⚠ 🔲
Photo: M. Göbel

S74134-4 *Symphysodon aequifasciatus* Pellegrin, 1904
DISCUS ROT-TÜRKIS / RED-TURQUOISE
Zuchtform/breeding form "Red Silk F_1"

♂ ⚠ ℙ ◐ ☺ 🖳 🐟 ⚠ 🔲
Photo: M. Göbel

S74135-4 *Symphysodon aequifasciatus* Pellegrin, 1904
DISCUS ROT-TÜRKIS / RED-TURQUOISE
Zuchtform/breeding form "Red Silk F_1"

♀ ⚠ ℙ ◐ ☺ 🖳 🐟 ⚠ 🔲
Photo: M. Göbel

S74104-4 *Symphysodon aequifasciatus* Pellegrin, 1904
DISCUS ROT-TÜRKIS / RED-TURQUOISE
Zuchtform/breeding form "Red Silk F_2"

♂ ⚠ ℙ ◐ ☺ 🖳 🐟 ⚠ 🔲
Photo: M. Göbel

S74136-4 *Symphysodon aequifasciatus* Pellegrin, 1904
DISCUS ROT-TÜRKIS / RED-TURQUOISE
Zuchtform/breeding form "Red Silk F_2"

♀ ⚠ ℙ ◐ ☺ 🖳 🐟 ⚠ 🔲
Photo: M. Göbel

Gruppe 7 / Group 7: Rot-türkis / Red-turquoise

Gruppe 7 / Group 7: Rot-türkis / Red-turquoise

S74105-4 *Symphysodon aequifasciatus* Pellegrin, 1904
DISCUS ROT-TÜRKIS / RED-TURQUOISE
Zuchtform/breeding form   "Red Silk F$_3$"
Photo: M. Göbel

S74106-4 *Symphysodon aequifasciatus* Pellegrin, 1904
DISCUS ROT-TÜRKIS / RED-TURQUOISE
Zuchtform/breeding form   "Red Silk F$_3$"
Photo: M. Göbel

S74107-4 *Symphysodon aequifasciatus* Pellegrin, 1904
DISCUS ROT-TÜRKIS / RED-TURQUOISE
Zuchtform/breeding form   "Red Silk F$_3$"
Photo: M. Göbel

S74108-4 *Symphysodon aequifasciatus* Pellegrin, 1904
DISCUS ROT-TÜRKIS / RED-TURQUOISE
Zuchtform/breeding form   "Red Silk F$_3$"
Photo: M. Göbel

S74109-4 *Symphysodon aequifasciatus* Pellegrin, 1904
DISCUS ROT-TÜRKIS / RED-TURQUOISE
Zuchtform/breeding form   "Red Silk F$_3$"
Photo: M. Göbel

S74110-4 *Symphysodon aequifasciatus* Pellegrin, 1904
DISCUS ROT-TÜRKIS / RED-TURQUOISE
Zuchtform/breeding form   "Red Silk F$_3$"
Photo: M. Göbel

S90101-4 *Symphysodon aequifasciatus* Pellegrin, 1904
DISCUS flächig rot (gelb/braun) / solid red (yellow/brown)
Zuchtform/breeding form "YELLOW"

Photo: Wayne DC Hkg.

S90102-4 *Symphysodon aequifasciatus* Pellegrin, 1904
DISCUS flächig rot (gelb/braun) / solid red (yellow/brown)
Zuchtform/breeding form "GOLDEN"

Photo: Wayne DC Hkg.

S90103-4 *Symphysodon aequifasciatus* Pellegrin, 1904
DISCUS flächig rot (gelb/braun) / solid red (yellow/brown)
Zuchtform/breeding form "GOLDEN"

Photo: Wayne DC Hkg.

S90104-4 *Symphysodon aequifasciatus* Pellegrin, 1904
DISCUS flächig rot (gelb/braun) / solid red (yellow/brown)
Zuchtform/breeding form "GOLDEN-RAINBOW"

Photo: Takrit-Aquarium Bangkok

S90105-4 *Symphysodon aequifasciatus* Pellegrin, 1904
DISCUS flächig rot (gelb/braun) / solid red (yellow/brown)
Zuchtform/breeding form "SWEETS TEMPTATION"

Photo: Wayne DC Hkg.

S90106-4 *Symphysodon aequifasciatus* Pellegrin, 1904
DISCUS flächig rot (gelb/braun) / solid red (yellow/brown)
Zuchtform/breeding form "GOLDEN-PHOENIX"

Photo: Godwin K.N.Sim

Gruppe 8 / Group 8: Discus flächig rot (gelb/braun) / Solid red (yellow/brown)

South American **Cichlids IV** — **163**

S74110-4 *Symphysodon aequifasciatus* Pellegrin, 1904
DISCUS ROT-TÜRKIS / RED-TURQUOISE
Zuchtform/breeding form "Red Silk F_3"

Photo: M. Göbel

S90107-4 *Symphysodon aequifasciatus* Pellegrin, 1904
DISCUS flächig rot (gelb/braun) / solid red (yellow/brown)
Zuchtform/breeding form "RED-ANGEL"
Photo: WayneDC Hkg.

S90108-4 *Symphysodon aequifasciatus* Pellegrin, 1904
DISCUS flächig rot (gelb/braun) / solid red (yellow/brown)
Zuchtform/breeding form "BLUE-ANGEL"
Photo: H.J. Mayland

S90109-4 *Symphysodon aequifasciatus* Pellegrin, 1904
DISCUS flächig rot (gelb/braun) / solid red (yellow/brown)
Zuchtform/breeding form "RED STRIPEHEAD"
Photo: M. Göbel

S90110-4 *Symphysodon aequifasciatus* Pellegrin, 1904
DISCUS flächig rot (gelb/braun) / solid red (yellow/brown)
Zuchtform/breeding form "MARLBORO-RED"
Photo: Takrit-Aquarium Bangkok

S90111-4 *Symphysodon aequifasciatus* Pellegrin, 1904
DISCUS flächig rot (gelb/braun) / solid red (yellow/brown)
Zuchtform/breeding form "TOMATO-RED"
Photo: Godwin K.N. Sim

S90112-4 *Symphysodon aequifasciatus* Pellegrin, 1904
DISCUS flächig rot (gelb/braun) / solid red (yellow/brown)
Zuchtform/breeding form "MARLBORO-RED"
Photo: Takrit-Aquarium Bangkok

Gruppe 8 / Group 8: Discus flächig rot (gelb/braun) / Solid red (yellow/brown)

Gruppe 8 / Group 8: Discus flächig rot (gelb/braun) / Solid red (yellow/brown)

S90113-4 *Symphysodon aequifasciatus* Pellegrin, 1904
DISCUS flächig rot (gelb/braun) / solid red (yellow/brown)
Zuchtform/breeding form "PIDGEON BLOOD"

Photo: H.J. Mayland

S90114-4 *Symphysodon aequifasciatus* Pellegrin, 1904
DISCUS flächig rot (gelb/braun) / solid red (yellow/brown)
Zuchtform/breeding form „Red Silk" F_3

Photo: M. Göbel

S89412-4 *Symphysodon aequifasciatus* Pellegrin, 1904
DISCUS flächig rot (gelb/braun) / solid red (yellow/brown)
Brazil, wild

Photo: M. Göbel

S74132-4 *Symphysodon aequifasciatus* Pellegrin, 1904
DISCUS flächig rot (gelb/braun) / solid red (yellow/brown)
Zuchtform/breeding form „Tangerine Dream" F_3

Photo: M. Göbel

S90117-4 *Symphysodon aequifasciatus* Pellegrin, 1904
DISCUS flächig rot/red with stripes
Brazil, wild

Photo: H.J. Mayland

S90214-4 *Symphysodon aequifasciatus* Pellegrin, 1904
DISCUS flächig rot/red with stripes
Zuchtform/breeding form „Alenquer, Red Eddy"

Photo: M. Göbel

S90118-4 *Symphysodon aequifasciatus* PELLEGRIN, 1904
DISCUS flächig rot (gelb/braun) / solid red (yellow/brown)
Brazil, wild „Rio Madeira" (mother of S90119)

♀ ⚠ ℙ ◐ ☺ 🔲 🔳 ➤ ⚠ 🔲   Photo: M. Göbel

S90119-4 *Symphysodon aequifasciatus* PELLEGRIN, 1904
DISCUS flächig rot (gelb/braun) / solid red (yellow/brown)
Zuchtform/breeding form „Rio Madeira" F_1

⚠ ℙ ◐ ☺ 🔲 🔳 ➤ ⚠ 🔲   Photo: M. Göbel

Gruppe 8 / Group 8: Discus flächig rot (gelb/braun) / Solid red (yellow/brown)

S90100-4 *Symphysodon aequifasciatus* PELLEGRIN, 1904
DISCUS flächig rot (gelb/braun) / solid red (yellow/brown)
Zuchtform/breeding form

⚠ ℙ ◐ ☺ 🔲 🔳 ➤ ⚠ 🔲   Photo: M. Göbel

Gruppe 9 / Group 9: Offene Klasse / Open Class

S75001-4 *Symphysodon aequifasciatus* Pellegrin, 1904
DISCUS "ROT-GEPERLT / RED-PEARL"
Zuchtform/breeding form

Photo: Wayne DC Hkg.

S75002-4 *Symphysodon aequifasciatus* Pellegrin, 1904
DISCUS "ROT-GEPERLT / RED-PEARL"
Zuchtform/breeding form

Photo: Archiv A.C.S.

S75045-4 *Symphysodon aequifasciatus* Pellegrin, 1904
DISCUS "CHECKERBOARD"
Zuchtform/breeding form

Photo: M. Göbel

S75060-4 *Symphysodon aequifasciatus* Pellegrin, 1904
DISCUS "CHECKERBOARD"
Zuchtform/breeding form

Photo: M. Göbel

S75061-4 *Symphysodon aequifasciatus* Pellegrin, 1904
DISCUS "CHECKERBOARD"
Zuchtform/breeding form

Photo: M. Göbel

S75062-4 *Symphysodon aequifasciatus* Pellegrin, 1904
DISCUS "CHECKERBOARD"
Zuchtform/breeding form

Photo: M. Göbel

S75003-4 *Symphysodon aequifasciatus* Pellegrin, 1904
DISCUS "GIANT PEARL"
Zuchtform/breeding form

Photo: Wayne DC HongKong

S75004-4 *Symphysodon aequifasciatus* Pellegrin, 1904
DISCUS "JEWEL OF MAYANS"
Zuchtform/breeding form

Photo: Wayne DC HongKong

S91516-4 *Symphysodon aequifasciatus* Pellegrin, 1904
DISCUS "RED SPOTTED"
Zuchtform/breeding form

Photo: M. Göbel

S75006-4 *Symphysodon aequifasciatus* Pellegrin, 1904
DISCUS "BLUE JEWEL"
Zuchtform/breeding form

Photo: H.J. Mayland

S75007-4 *Symphysodon aequifasciatus* Pellegrin, 1904
DISCUS "BLUE PEARL JEWEL"
Zuchtform/breeding form

Photo: Lee Tong Juan SI

S75008-4 *Symphysodon aequifasciatus* Pellegrin, 1904
DISCUS "RED SPOTTED JEWEL"
Zuchtform/breeding form

Photo: Lee Tong Juan SI

Gruppe 9 / Group 9: Offene Klasse / Open Class

S75009-4 *Symphysodon aequifasciatus* PELLEGRIN, 1904
DISCUS "CHOCOLATE PEARL", Breeding form
Champion „open class" -AQUARAMA 1995-

Photo: Godwin K.N. Sim

S75010-4 *Symphysodon aequifasciatus* PELLEGRIN, 1904
DISCUS "GODWIN SIM"
Zuchtform/breeding form

Photo: Godwin K.N. Sim

S75011-4 *Symphysodon aequifasciatus* PELLEGRIN, 1904
DISCUS "BLOOD PEARL"
Zuchtform/breeding form

Photo: Godwin K.N. Sim

S75012-4 *Symphysodon aequifasciatus* PELLEGRIN, 1904
DISCUS "BLOOD PEARL -big spots-"
Zuchtform/breeding form

Photo: Godwin K.N. Sim

S75013-4 *Symphysodon aequifasciatus* PELLEGRIN, 1904
DISCUS "BLOOD PEARL -red body-"
Zuchtform/breeding form

Photo: Godwin K.N. Sim

S75013-4 *Symphysodon aequifasciatus* PELLEGRIN, 1904
DISCUS "BLOOD PEARL -red body-"
Zuchtform/breeding form

Photo: Godwin K.N. Sim

S75015-4 *Symphysodon aequifasciatus* Pellegrin, 1904
DISCUS "PEARL HIGHFIN"
Zuchtform/breeding form

Photo: H.J. Mayland

S75016-4 *Symphysodon aequifasciatus* Pellegrin, 1904
DISCUS "STRIPED HIGHFIN"
Zuchtform/breeding form

Photo: H.J. Mayland

S75047-4 *Symphysodon aequifasciatus* Pellegrin, 1904
DISCUS "GLASSFIN"
Zuchtform/breeding form

Photo: ACS/TFF Mainland

S75046-4 *Symphysodon aequifasciatus* Pellegrin, 1904
DISCUS "SNAKE SKIN"
Zuchtform/breeding form

Photo: Archiv A.C.S./Tomizana

S75048-4 *Symphysodon aequifasciatus* Pellegrin, 1904
DISCUS "SNAKE SKIN"
Zuchtform/breeding form

Photo: Lee Tong Juan Sl..

S75049-4 *Symphysodon aequifasciatus* Pellegrin, 1904
DISCUS "SNAKE SKIN"
Zuchtform/breeding form

Photo: Lee Tong Juan, Sl.

Gruppe 9 / Group 9: Offene Klasse / Open Class

Gruppe 9 / Group 9: Offene Klasse / Open Class

S75017-4 *Symphysodon aequifasciatus* Pellegrin, 1904
DISCUS "offene Klasse / open class"
Zuchtform/breeding form

Photo: H.J. Mayland

S75018-4 *Symphysodon aequifasciatus* Pellegrin, 1904
DISCUS "offene Klasse / open class"
Zuchtform/breeding form

Photo: H.J. Mayland

S75019-4 *Symphysodon aequifasciatus* Pellegrin, 1904
DISCUS "offene Klasse / open class"
Zuchtform/breeding form

Photo: H.J. Mayland

S75020-4 *Symphysodon aequifasciatus* Pellegrin, 1904
DISCUS "offene Klasse / open class"
Zuchtform/breeding form

Photo: H.J. Mayland

S75021-4 *Symphysodon aequifasciatus* Pellegrin, 1904
DISCUS "offene Klasse / open class"
Zuchtform/breeding form

Photo: H.J. Mayland

S75022-4 *Symphysodon aequifasciatus* Pellegrin, 1904
DISCUS "offene Klasse / open class"
Zuchtform/breeding form

Photo: H.J. Mayland

South American **Cichlids IV**

© Verlag A.C.S. GmbH

S75023-3 *Symphysodon aequifasciatus* Pellegrin, 1904
DISCUS "offene Klasse / open class" „GHOST"
Zuchtform/breeding form

Photo: ACS/TFF Mainland

S75025-3 *Symphysodon aequifasciatus* Pellegrin, 1904
DISCUS "offene Klasse / open class" "GHOST"
Zuchtform/breeding form

Photo: H.J. Mayland

S75024-3 *Symphysodon aequifasciatus* Pellegrin, 1904
DISCUS "offene Klasse / open class"
Zuchtform/breeding form

Photo: H.J. Mayland

S75050-4 *Symphysodon aequifasciatus* Pellegrin, 1904
DISCUS "offene Klasse / open class"
Zuchtform/breeding form

Photo: H.J. Mayland

S75051-4 *Symphysodon aequifasciatus* Pellegrin, 1904
DISCUS "offene Klasse / open class"
Zuchtform/breeding form

Photo: H.J. Mayland

S75052-4 *Symphysodon aequifasciatus* Pellegrin, 1904
DISCUS "offene Klasse / open class"
Zuchtform/breeding form

Photo: H.J. Mayland

Gruppe 9 / Group 9: Offene Klasse / Open Class

S75064-4 *Symphysodon aequifasciatus* PELLEGRIN, 1904
DISCUS "offene Klasse / open class"
Zuchtform/breeding form   "GOLDEN GHOST"

Photo: A.C.S./Tomizana

S75063-3 *Symphysodon aequifasciatus* PELLEGRIN, 1904
DISCUS "offene Klasse / open class"
Zuchtform/breeding form   "GOLDEN GHOST"

Photo: H.J. Mayland

S75026-4 *Symphysodon aequifasciatus* PELLEGRIN, 1904
DISCUS "offene Klasse / open class"
Zuchtform/breeding form

Photo: H.J. Mayland

S75053-4 *Symphysodon aequifasciatus* PELLEGRIN, 1904
DISCUS "offene Klasse / open class"
Zuchtform/breeding form   "YELLOW RAINBOW"

Photo: Sea View Aquarium

S75053-2 *Symphysodon aequifasciatus* PELLEGRIN, 1904
DISCUS "offene Klasse / open class" JUVENIL
Zuchtform/breeding form   "YELLOW RAINBOW"

Photo: Sea View Aquarium

S75054-4 *Symphysodon aequifasciatus* PELLEGRIN, 1904
DISCUS "offene Klasse / open class"
Zuchtform/breeding form   "SKYBLUE"

Photo: W. Mikschofsky

S75027-4 *Symphysodon aequifasciatus* Pellegrin, 1904
DISCUS "offene Klasse / open class"
Zuchtform/breeding form "PIDGEON BLOOD"

Photo: M. Göbel

S75027-4 *Symphysodon aequifasciatus* Pellegrin, 1904
DISCUS "offene Klasse / open class"
Zuchtform/breeding form "PIDGEON BLOOD"

Photo: M. Göbel

S75029-4 *Symphysodon aequifasciatus* Pellegrin, 1904
DISCUS "offene Klasse / open class"
Zuchtform/breeding form "PIDGEON BLOOD"

Photo: M. Göbel

S75030-4 *Symphysodon aequifasciatus* Pellegrin, 1904
DISCUS "offene Klasse / open class"
Zuchtform/breeding form "PIDGEON BLOOD"

Photo: M. Göbel

S75031-4 *Symphysodon aequifasciatus* Pellegrin, 1904
DISCUS "offene Klasse / open class"
Zuchtform/breeding form "PIDGEON BLOOD"

Photo: H.J. Mayland

S75032-4 *Symphysodon aequifasciatus* Pellegrin, 1904
DISCUS "offene Klasse / open class"
Zuchtform/breeding form "PIDGEON BLOOD"

Photo: H.J. Mayland

Gruppe 9 / Group 9: Offene Klasse / Open Class

Gruppe 9 / Group 9: Offene Klasse / Open Class

S75033-4 *Symphysodon aequifasciatus* Pellegrin, 1904
DISCUS "offene Klasse / open class"
Zuchtform/breeding form "PIDGEON BLOOD"

Photo: H.J. Mayland

S75034-4 *Symphysodon aequifasciatus* Pellegrin, 1904
DISCUS "offene Klasse / open class"
Zuchtform/breeding form "PIDGEON BLOOD"

Photo: H.J. Mayland

S75035-4 *Symphysodon aequifasciatus* Pellegrin, 1904
DISCUS "offene Klasse / open class"
Zuchtform/breeding form "PIDGEON BLOOD"

Photo: H.J. Mayland

S75036-4 *Symphysodon aequifasciatus* Pellegrin, 1904
DISCUS "offene Klasse / open class"
Zuchtform/breeding form "PIDGEON BLOOD"

Photo: H.J. Mayland

S75037-4 *Symphysodon aequifasciatus* Pellegrin, 1904
DISCUS "offene Klasse / open class"
Zuchtform/breeding form "PIDGEON BLOOD"

Photo: H.J. Mayland

S75038-4 *Symphysodon aequifasciatus* Pellegrin, 1904
DISCUS "offene Klasse / open class"
Zuchtform/breeding form "PIDGEON BLOOD"

Photo: H.J. Mayland

South American **Cichlids** *IV* © Verlag A.C.S. GmbH

S75039-4 *Symphysodon aequifasciatus* Pellegrin, 1904
DISCUS "offene Klasse / open class"
Zuchtform/breeding form "BLUE SAPHIRE"

Photo: Godwin K.N. Sim

S75040-4 *Symphysodon aequifasciatus* Pellegrin, 1904
DISCUS "offene Klasse / open class"
Zuchtform/breeding form "PIDGEON BLOOD"

Photo: A.C.S./Tomizana

S75041-4 *Symphysodon aequifasciatus* Pellegrin, 1904
DISCUS "offene Klasse / open class"
Zuchtform/breeding form "PIDGEON BLOOD"

Photo: H.J. Mayland

S75042-4 *Symphysodon aequifasciatus* Pellegrin, 1904
DISCUS "offene Klasse / open class"
Zuchtform/breeding form "PIDGEON BLOOD"

Photo: A.C.S./Tomizana

S75043-4 *Symphysodon aequifasciatus* Pellegrin, 1904
DISCUS "offene Klasse / open class"
Zuchtform/breeding form "PIDGEON BLOOD"

Photo: H.J. Mayland

S75044-4 *Symphysodon aequifasciatus* Pellegrin, 1904
DISCUS "offene Klasse / open class"
Zuchtform/breeding form "PIDGEON BLOOD"

Photo: H. Morche

S75055-4 *Symphysodon aequifasciatus* Pellegrin, 1904
DISCUS "offene Klasse / open class"
Zuchtform/breeding form „PIDGEON BLOOD"

Photo: M. Smith

Im Biotop des Hohen Skalars, *Pterophyllum altum*, hat das ansonsten glasklare Wasser eine tiefbraune Tönung.

In the biotope of the Real Altum Angel, Pterophyllum altum, *the water is both deep brown and cristal clear*

Photo: U. Werner

Photo: H.J. Mayland

Photo: A. Canovas

Der Hohe Skalar - die begehrteste Wildform der Segelflosser in der Aquaristik.

The Real Altum Angel - the most desired wild species of angels in the hobby.

S71925-1 *Pterophyllum* cf. *scalare* (see also p. 161 ff.)
sogenannter / so-called "Peru-altum", Baby
Peru (Rio Ucayali area?), W, 20 cm

Photo: F. Teigler / A.C.S.

S71925-2 *Pterophyllum* cf. *scalare* (see also p. 161 ff.)
sogenannter / so-called "Peru-altum", juvenil
Peru (Rio Ucayali area?), W, 20 cm

Photo: F. Teigler / A.C.S.

S71900-1 *Pterophyllum altum* PELLEGRIN, 1903
Hoher Segelflosser / Real Altum Angel
Orinoco, W, 25 cm

Photo: F. Teigler / A.C.S.

S71900-2 *Pterophyllum altum* PELLEGRIN, 1903
Hoher Segelflosser / Real Altum Angel
Orinoco, W, 25 cm

Photo: F. Teigler / A.C.S.

S71900-3 *Pterophyllum altum* PELLEGRIN, 1903
Hoher Segelflosser / Real Altum Angel
Orinoco, W, 25 cm

Photo: Migge-Reinhardt / A.C.S.

S71900-4 *Pterophyllum altum* PELLEGRIN, 1903
Hoher Segelflosser / Real Altum Angel
Orinoco, W, 25 cm

Photo: H. J. Mayland

Skalare / Angels: Wildformen / Wild forms

S71900-5 *Pterophyllum altum* Pellegrin, 1903
Hoher Segelflosser / Real Altum Angel
Orinoco, W, 25 cm

Photo: H. Morche

S71900-5 *Pterophyllum altum* Pellegrin, 1903
Hoher Segelflosser / Real Altum Angel
Orinoco, W, 25 cm

Photo: H. J. Mayland

S71901-4 *Pterophyllum altum* Pellegrin, 1903
Hoher Segelflosser / Real Altum Angel
Rio Negro, W, 25 cm.

Photo: Schmidt-Knaatz

S71905-4 *Pterophyllum altum* Pellegrin, 1903
Hoher Segelflosser / Real Altum Angel
F_1, B, 25 cm

Photo: H. Linke

S71907-3 *Pterophyllum altum* Pellegrin, 1903
Hoher Segelflosser / Real Altum Angel
F_2, B, 25 cm

Photo: H. Linke

S71907-3 *Pterophyllum altum* Pellegrin, 1903
Hoher Segelflosser / Real Altum Angel
F_2, B, 25 cm

Photo: H. Linke

Photos: H. Linke

1. Nachzuchttiere von *Pterophyllum altum*, 60 Tage alt und ca. 9 cm Spannweite

Born in captivity: 60 days old Pterophyllum altum, about 9 cm high.

2. Im Alter von 20 Tagen sind die Tiere etwa 4 cm hoch.

The same fish 20 days old. At this age they are about 4 cm high.

Skalare / Angels: Wildformen / Wild forms

S71910-3 *Pterophyllum* cf. *leopoldi*
Leopolds Skalar / Leopold´s Angel
WF-NZ,unbek.Herk./origin unknown, B, 15 cm

Photo: F. Teigler /A.C.S.

S71912-3 *Pterophyllum leopoldi* (GOSSE, 1963)
Leopolds Skalar / Leopold´s Angel
Import via Iquitos / Peru, W, 15 cm

Photo: H. J. Mayland

S71912-4 *Pterophyllum leopoldi* (GOSSE, 1963)
Leopolds Skalar / Leopold´s Angel
Import via Iquitos / Peru, W, 15 cm

Photo: H. J. Mayland

S71912-4 *Pterophyllum leopoldi* (GOSSE, 1963)
Leopolds Skalar / Leopold´s Angel
Import via Iquitos / Peru, W, 15 cm

Photo: H. J. Mayland

S71912-5 *Pterophyllum leopoldi* (GOSSE, 1963)
Leopolds Skalar / Leopold´s Angel
Import via Iquitos / Peru, W, 15 cm

Photo: H. J. Mayland

S71912-5 *Pterophyllum leopoldi* (GOSSE, 1963)
Leopolds Skalar / Leopold´s Angel
Import via Iquitos / Peru, W, 15 cm

Photo: H. J. Mayland

South American Cichlids IV © Verlag A.C.S. GmbH

Photos: H. J. Mayland

Dieses Paar von *Pterophyllum leopoldi* laicht gerade an der Frontscheibe des Aquariums ab. Das fertige Gelege wird, wie bei allen Skalaren, von beiden Elterntieren gemeinsam bewacht

This pair of* Pterophyllum leopoldi *in the process of spawning on the front pane. As in all angels, both parents are looking after eggs and young.

Skalare / Angels: Wildformen / Wild forms

S71935-4 *Pterophyllum scalare* (LICHTENSTEIN, 1823)
Skalar / Angel
Import via Manaus/Brazil, W, 15 cm
Photo: H. Linke

S71935-4 *Pterophyllum scalare* (LICHTENSTEIN, 1823)
Skalar / Angel
Import via Manaus/Brazil, W, 15 cm
Photo: H. Linke

S71936-5 *Pterophyllum scalare* (LICHTENSTEIN, 1823)
Skalar / Angel
Fresh import via Manaus/Brazil, W, 15 cm
Photo: F. Teigler / A.C.S.

S71936-5 *Pterophyllum scalare* (LICHTENSTEIN, 1823)
Skalar / Angel
Fresh import via Manaus/Brazil, W, 15 cm
Photo: F. Teigler / A.C.S.

S71937-5 *Pterophyllum scalare* (LICHTENSTEIN, 1823)
Skalar / Angel
Brazil, W, 15 cm
Photo: H. Morche

S71937-5 *Pterophyllum scalare* (LICHTENSTEIN, 1823)
Skalar / Angel   Portrait
Brazil, W, 15 cm
Photo: Migge-Reinhard / A.C.S.

S71939-4 *Pterophyllum scalare* (LICHTENSTEIN, 1823)
Skalar / Angel (sogenannter / so-called "eimekei")
Brazil, W, 15 cm

Photo: F. Teigler / A.C.S.

S71940-5 *Pterophyllum scalare* (LICHTENSTEIN, 1823)
Skalar / Angel (sogenannter / so-called "eimekei")
Aquarienstamm / Aquarium strain, B, 15 cm

Photo: J. Glaser

S71940-5 *Pterophyllum scalare* (LICHTENSTEIN, 1823)
Skalar / Angel (sogenannter / so-called "eimekei")
Aquarienstamm / Aquarium strain, B, 15 cm

Photo: J. Glaser

S71940-5 *Pterophyllum scalare* (LICHTENSTEIN, 1823)
Skalar / Angel (sogenannter / so-called "eimekei")
Aquarienstamm / Aquarium strain, B, 15 cm

Photo: J. Glaser

S71940-5 *Pterophyllum scalare* (LICHTENSTEIN, 1823)
Skalar / Angel (sogenannter / so-called "eimekei")
Aquarienstamm / Aquarium strain, B, 15 cm

Photo: H. J. Mayland

S71940-4 *Pterophyllum scalare* (LICHTENSTEIN, 1823)
Skalar / Angel (sogenannter / so-called "eimekei")
Aquarienstamm / Aquarium strain, B, 15 cmt

Photo: Nakano / A.C.S.

Skalare / Angels: Wildformen / Wild forms

S71945-3 *Pterophyllum scalare* (Lichtenstein, 1823)
Skalar / Angel „Red Spotted"
Import from Guyana, W, 15 cm

Photo: H. Linke

S71945-4 *Pterophyllum scalare* (Lichtenstein, 1823)
Skalar / Angel „Red Spotted"
Import from Guyana, W, 15 cm

Photo: Nakano / A.C.S.

S71947-5 *Pterophyllum scalare* (Lichtenstein, 1823)
Skalar / Angel „Red Spotted"
Import from Peru, W, 15 cm

Photo: F. Teigler / A.C.S.

S71947-5 *Pterophyllum scalare* (Lichtenstein, 1823)
Skalar / Angel „Red Spotted"
Import from Peru, W, 15 cm

Photo: F. Teigler / A.C.S.

S71947-5 *Pterophyllum scalare* (Lichtenstein, 1823)
Skalar / Angel „Red Spotted"
Import from Peru, W, 15 cm

Photo: F. Teigler / A.C.S.

S71947-4 *Pterophyllum scalare* (Lichtenstein, 1823)
Skalar / Angel „Red Spotted"
Import from Peru, W, 15 cm

Photo: F. Teigler / A.C.S.

South American Cichlids IV   © Verlag A.C.S. GmbH

S71947-5 *Pterophyllum scalare* (LICHTENSTEIN, 1823)
Skalar / Angel „Red Spotted"
Import fromPeru, W, 15 cm

Photo: F. Teigler / A.C.S.

S71926-1 *Pterophyllum* cf. *scalare* (see also p. 153)
sogenannter / so-called "Peru-altum"
Peru (Rio Ucayali area?), W, 20 cm

Photo: F. Teigler / A.C.S.

S71927-2 *Pterophyllum* cf. *scalare* (see also p. 153)
sogenannter / so-called "Peru-altum"
Peru (Rio Ucayali area?), W, 20 cm

Photo: F. Schäfer

S71927-4 *Pterophyllum* cf. *scalare* (see also p. 153)
sogenannter / so-called "Peru-altum"
Peru (Rio Ucayali area?), W, 20 cm

Photo: F. Teigler / A.C.S.

S71928-4 *Pterophyllum* cf. *scalare* (see also p. 153)
sogenannter / so-called "Peru-altum", spotted
Peru (Rio Ucayali area?), W, 20 cm

Photo: H. Linke

S71928-4 *Pterophyllum* cf. *scalare* (see also p. 153)
sogenannter / so-called "Peru-altum", spotted
Peru (Rio Ucayali area?), W, 20 cm

Photo: H. Linke

Skalare / Angels: Wildformen / Wild forms

S71930-4 *Pterophyllum* cf. *scalare* (see also p. 153)
sogenannter / so-called "Peru-altum", Var. II
Peru (Rio Amazon area?), W, 20 cm

Photo: H. J. Mayland

S71930-4 *Pterophyllum* cf. *scalare* (see also p. 153)
sogenannter / so-called "Peru-altum", Var. II
Peru (Rio Amazon area?), W, 20 cm

Photo H. J. Mayland

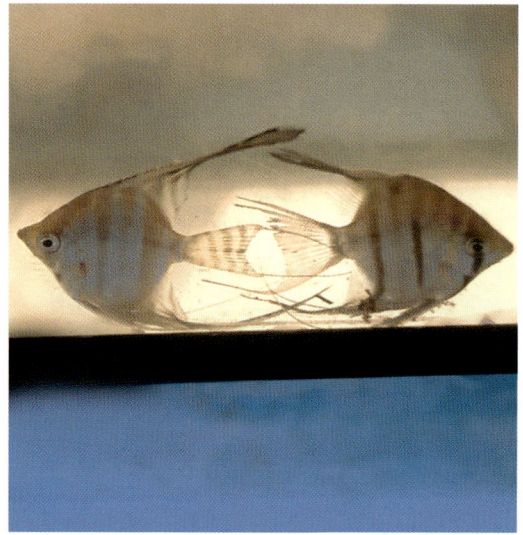

S71932-4 *Pterophyllum* sp. (two species?)
caught together in the Rio Negro (leg. DEN DAAS)
Brazil: Rio Negro, W, 15 cm

Photo: I. den Daas

S71933-5 *Pterophyllum* sp. (two species?)
caught together in the Rio Arian (leg. DEN DAAS)
Brazil: Rio Negro area, W, 15 cm

Photo: I. den Daas

S71934-3 *Pterophyllum* sp.
caught in the Rio Cuiuni (leg. DAWES)
Brazil: Rio Negro area, W, 15 cm

Photo: J. Dawes

S71916-3 *Pterophyllum* sp. (so-called „dumerilii")
Schafskopf-Skalar / Sheepshead Angel
Brazil, W, 15 cm

Photo: F. Teigler / A.C.S.

S71917-3 *Pterophyllum* sp. (so-called „dumerilii")
Schafskopf-Skalar / Sheepshead Angel
Brazil, W, 15 cmt

Photo: F. Teigler / A.C.S.

S71918-3 *Pterophyllum* sp. (so-called „dumerilii")
Schafskopf-Skalar / Sheepshead Angel
Brazil, W, 15 cm

Photo: F. Teigler / A.C.S.

S71919-3 *Pterophyllum* sp. (so-called „dumerilii")
Schafskopf-Skalar / Sheepshead Angel
Brazil, W, 15 cm

Photo: F. Teigler / A.C.S.

S71920-3 *Pterophyllum* sp. (so-called „dumerilii")
Schafskopf-Skalar / Sheepshead Angel
Brazil, W, 15 cm

Photo: E. Schraml / A.C.S.

S71921-3 *Pterophyllum* sp. (so-called „dumerilii")
Schafskopf-Skalar / Sheepshead Angel
Brazil, W, 15 cm

Photo: F. Teigler / A.C.S.

S71922-3 *Pterophyllum* sp. (so-called „dumerilii")
Schafskopf-Skalar / Sheepshead Angel
Brazil, W, 15 cm

Photo: F. Teigler / A.C.S.

Die Aufnahmen zeigen, wie unterschiedlich der Schafskopf-Skalar gezeichnet sein kann. In der Aquaristik wird die Art als *P. dumerilii* bezeichnet. Es handelt sich aber um eine wahrscheinlich noch unbeschriebene Art. *P. dumerilii* ist ein Synonym zu *P. scalare* (S. KULLANDER, 1986).

The pictures show how much the individual colouration differs in the Sheepshead Angel. This species is called Pterophyllum dumerilii *in the hobby. In all propability it is a new, undescribed species.* P. dumerilii *is a synonym of* P. scalare *(see KULLANDER, 1986).*

Photos: Nakano / A.C.S.

Die Zucht von Skalaren ist bei Aquarianers ebenso beliebt wie die Erzüchtung neuer Farbvarianten.

The breeding of angels - a hobby in the hobby. Both - breeding the fish and creating new varieties is enjoyed by the fishkeepers.

S72000-4 Segelflosser/Angelfish "Vierstreifen/Four-Stripes" Zucht-Variante/Breeding form

Photo: A. Canovas

S72005-3 Segelflosser / Angelfish
Hell, Normalflosser / Bright, normal finned
B, Z, 12-15cm

Photo: J. Glaser

S72006-3 Segelflosser / Angelfish
Hell, Schwarzflosser / Bright, black finned
B, Z, 12-15cm

Photo: J. Glaser

S72006-4 Segelflosser / Angelfish
Hell, Schwarzflosser / Bright, black finned
B, Z, 12-15cm

Photo: J. Glaser

S72005-2 Segelflosser / Angelfish
Hell, Normalflosser / Bright, normal finned
B, Z, 12-15cm

Photo: F. Teigler

S72006-3 Segelflosser / Angelfish
Hell, Schwarzflosser / Bright, black finned
B, Z, 12-15cm

Photo: J. Glaser

S72010-4 Segelflosser / Angelfish
Hell, Schleier / Bright, veiltail
B, Z, 12-15cm

Photo: H. Linke

Skalare / Angels: Zuchtformen / Breeding forms

South American Cichlids IV

Skalare / Angels: Zuchtformen / Breeding forms

S72005-4 Segelflosser / Angelfish
Hell, Normalflosser / Bright, normal finned
B, Z, 12-15cm

Photo: Nakano / A.C.S.

S72011-4 Segelflosser / Angelfish
Hell, Schleier / Bright, veiltail-hifin
B, Z, 12-15cm

Photo: Nakano / A.C.S.

S72012-3 Segelflosser / Angelfish
Hell, Hochflosser / Bright, hifin
B, Z, 12-15cm

Photo: F. Teigler

S72012-3 Segelflosser / Angelfish
Hell, Hochflosser / Bright, hifin
B, Z, 12-15cm

Photo: F. Teigler

S72005-4 Segelflosser / Angelfish
Hell, Normalflosser / Bright, normal finned
B, Z, 12-15cm

Photo: F. Teigler

S72005-3 Segelflosser / Angelfish
Hell, Normalflosser / Bright, normal finned
B, Z, 12-15cm

Photo: Nakano / A.C.S.

South American **Cichlids IV** © **Verlag A.C.S. GmbH**

S72013-2 Segelflosser / Angelfish
Rauchskalar, Normalflosser / Black Lace Angel
B, Z, 12-15cm
Photo: Nakano / A.C.S.

S72013-2 Segelflosser / Angelfish
Rauchskalar, Normalflosser / Black Lace Angel
B, Z, 12-15cm
Photo: F. Teigler

S72013-3 Segelflosser / Angelfish
Rauchskalar, Normalflosser / Black Lace Angel
B, Z, 12-15cm
Photo: F. Teigler

S72014-3 Segelflosser / Angelfish
Rauchskalar, Breitflosser / Black Lace Angel, broadfin
B, Z, 12-15cm
Photo: F. Teigler

S72014-3 Segelflosser / Angelfish
Rauchskalar, Breitflosser / Black Lace Angel, broadfin
B, Z, 12-15cm
Photo: F. Teigler

S72014-3 Segelflosser / Angelfish
Rauchskalar, Breitflosser / Black Lace Angel, broadfin
B, Z, 12-15cm
Photo: F. Teigler

South American **Cichlids IV**

Skalare / Angels: Zuchtformen / Breeding forms

S72015-3 Segelflosser / Angelfish
Schwarzstreifen / Black stripe
B, Z, 12-15cm

Photo: F. Teigler

S72015-4 Segelflosser / Angelfish
Schwarzstreifen / Black stripe
B, Z, 12-15cm

Photo: F. Teigler

S72015-3 Segelflosser / Angelfish
Schwarzstreifen / Black stripe
B, Z, 12-15cm

Photo: F. Teigler

S72016-3 Segelflosser / Angelfish
Schwarzstreifen, Schleier / Black stripe, veiltail
B, Z, 12-15cm

Photo: F. Teigler

S72015-3 Segelflosser / Angelfish
Schwarzstreifen / Black stripe
B, Z, 12-15cm

Photo: F. Teigler

S72015-3 Segelflosser / Angelfish
Schwarzstreifen / Black stripe
B, Z, 12-15cm

Photo: F. Teigler

S72011-4 Segelflosser / Angelfish
Hell, Schleier, Hochflosser / Bright, veiltail, hifin
B, Z, 12-15cm

Photo: F. Teigler

S72007-3 Segelflosser / Angelfish
Zebra, hell / Zebra-Angel, bright
B, Z, 12-15cm

Photo: F. Teigler

S72008-4 Segelflosser / Angelfish
Zebra, hell / Zebra-Angel, bright, hifin
B, Z, 12-15cm

Photo: F. Teigler

S72008-3 Segelflosser / Angelfish
Zebra, hell / Zebra-Angel, bright, hifin
B, Z, 12-15cm

Photo: F. Teigler

S72012-3 Segelflosser / Angelfish
Hell, Hochflosser / Bright, hifin
B, Z, 12-15cm

Photo: F. Teigler

S72010-4 Segelflosser / Angelfish
Hell, Schleier / Bright, veiltail
B, Z, 12-15cm

Photo: F. Teigler

Skalare / Angels: Zuchtformen / Breeding forms

South American Cichlids IV

Skalare / Angels: Zuchtformen / Breeding forms

S72030-3 Segelflosser / Angelfish
Zebra, Normalflosser / Zebra, normal finned
B, Z, 12-15cm

Photo: F. Teigler

S72035-2 Segelflosser / Angelfish
Leopard, Normalflosser / Leopard normal finned
B, Z, 12-15cm

Photo: F. Teigler

S72031-3 Segelflosser / Angelfish
Rauchskalar, Zebra / Black Lace Angel, zebra
B, Z, 12-15cm

Photo: Nakano / A.C.S.

S72032-4 Segelflosser / Angelfish
Rauchskalar, Zebra, Schleier / Black Lace Angel,
zebra, veiltail B, Z, 12-15cm

Photo: Nakano / A.C.S.

S72031-5 Segelflosser / Angelfish
Rauchskalar, Zebra / Black Lace Angel, zebra
B, Z, 12-15cm

Photo: H. Linke

S72031-4 Segelflosser / Angelfish
Rauchskalar, Zebra / Black Lace Angel, zebra
B, Z, 12-15cm

Photo: H. Linke

South American Cichlids IV © Verlag A.C.S. GmbH

S72040-2 Segelflosser / Angelfish
Halbschwarz / black-white, „Bicolor"
B, Z, 12-15cm
Photo: F. Teigler

S72040-4 Segelflosser / Angelfish
Halbschwarz / black-white, „Bicolor"
B, Z, 12-15cm
Photo: J. Glaser

S72040-5 Segelflosser / Angelfish
Halbschwarz / black-white, „Bicolor"
B, Z, 12-15cm
Photo: H. Linke

S72041-3 Segelflosser / Angelfish
Halbschwarz, Marmor / black-white, marbled
B, Z, 12-15cm
Photo: Nakano / A.C.S.

S72050-3 Segelflosser / Angelfish
Halbschwarz, Rotwange / black-white, red-wedge
B, Z, 12-15cm
Photo: Tomizana / A.C.S.

S72042-4 Segelflosser / Angelfish
Halbschwarz, Gold / black-white, gold
B, Z, 12-15cm
Photo: Nakano / A.C.S.

Skalare / Angels: Zuchtformen / Breeding forms

S72060-3 Segelflosser / Angelfish
Schwarz, Normalflosser / Black, normal finned
B, Z, 12-15cm

Photo: F. Teigler

S72060-4 Segelflosser / Angelfish
Schwarz, Normalflosser / Black, normal finned
B, Z, 12-15cm

Photo: Nakano / A.C.S.

S72060-3 Segelflosser / Angelfish
Schwarz, Normalflosser / Black, normal finned
B, Z, 12-15cm

Photo: Nakano / A.C.S.

S72060-4 Segelflosser / Angelfish
Schwarz, Normalflosser / Black, normal finned
B, Z, 12-15cm

Photo: Nakano / A.C.S.

S72061-3 Segelflosser / Angelfish
Schwarz, Breitflosser / Black, broad finned
B, Z, 12-15cm

Photo: F. Teigler

S72062-3 Segelflosser / Angelfish
Schwarz, Schleier / Black, veiltail
B, Z, 12-15cm

Photo: F. Teigler

South American Cichlids IV © Verlag A.C.S. GmbH

S72051-4 Segelflosser / Angelfish
Gold, Marmor / Golden, marbled
B, Z, 12-15cm

Photo: U. Werner

S72052-4 Segelflosser / Angelfish
Goldkopf, Marmor / Goldhead, marbled
"FALB - SCALAR", B, Z, 12-15cm

Photo: H. Linke

S72063-3 Segelflosser / Angelfish
Schwarz, Gold-Marmor / Black, Gold-Marble
B, Z, 12-15cm

Photo: F. Teigler

S72064-3 Segelflosser / Angelfish
Schwarz, Gold-Marmor, Schleier / Black, Gold-Marble, veiltail B, Z, 12-15cm

Photo: F. Teigler

S72051-3 Segelflosser / Angelfish
Gold-Marmor / Golden Marble
B, Z, 12-15cm

Photo: Nakano / A.C.S.

S72051-3 Segelflosser / Angelfish
Gold-Marmor / Golden Marble
B, Z, 12-15cm

Photo: J. Glaser

South American **Cichlids IV**

Skalare / Angels: Zuchtformen / Breeding forms

S72054-3 Segelflosser / Angelfish
Gold-Marmor, Hochflosser / Gold-Marble, hifin
B, Z, 12-15cm

Photo: F. Teigler

S72054-3 Segelflosser / Angelfish
Gold-Marmor, Hochflosser / Gold-Marble, hifin
B, Z, 12-15cm

Photo: F. Teigler

S72051-4 Segelflosser / Angelfish
Gold-Marmor / Gold-Marble
B, Z, 12-15cm

Photo: J. Glaser

S72051-3 Segelflosser / Angelfish
Gold-Marmor / Gold-Marble
B, Z, 12-15cm

Photo: J. Glaser

S72053-4 Segelflosser / Angelfish
Gold-Marmor, Schleier / Gold-Marble, veiltail
B, Z, 12-15cm

Photo: J. Glaser

S72053-3 Segelflosser / Angelfish
Gold-Marmor, Schleier / Gold-Marble, veiltail
B, Z, 12-15cm

Photo: F. Teigler

1. **S72056-4 Segelflosser / Angelfish**
   Goldkopf, Schleier
   Goldhead, veiltail

2. **S72057-5 Segelflosser / Angelfish**
   Schwarz, Gold-Marmor, Schleier
   Black, Gold-Marble, veiltail

Photos: A. Canovas

Skalare / Angels: Zuchtformen / Breeding forms

S72035-3 Segelflosser / Angelfish
Leopard
B, Z, 12-15cm
Photo: Nakano / A.C.S.

S72036-4 Segelflosser / Angelfish
Leopard, Schleier/Leopard, veiltail
B, Z, 12-15cm
Photo: Nakano / A.C.S.

S72036-4 Segelflosser / Angelfish
Leopard, Schleier / Leopard, veiltail
B, Z, 12-15cm
Photo: F. Teigler

S72036-4 Segelflosser / Angelfish
Leopard, Schleier / Leopard, veiltail
B, Z, 12-15cm
Photo: F. Teigler

S72037-4 Segelflosser / Angelfish
Leopard, Goldkopf, Schleier/
Leopard, Goldhead veiltail, B, Z, 12-15cm
Photo: F. Teigler

S72038-4 Segelflosser / Angelfish
Leopard, Goldkopf / Leopard, Goldhead
B, Z, 12-15cm
Photo: J. Glaser

206 *South American Cichlids IV* © **Verlag A.C.S. GmbH**

Skalare / Angels: Zuchtformen / Breeding forms

S72065-5 Segelflosser / Angelfish
Rotkopf, Marmor / Redhead, marbled
B, Z, 12-15cm
Photo: H. Linke

S72065-5 Segelflosser / Angelfish
Rotkopf, Marmor / Redhead, marbled
B, Z, 12-15cm
Photo: H. Linke

S72065-5 Segelflosser / Angelfish
Rotkopf, Marmor / Redhead, marbled
B, Z, 12-15cm, -- Befruchtung/insemination--
Photo: H. Linke

S72065-5 Segelflosser / Angelfish
Rotkopf, Marmor / Redhead, marbled
B, Z, 12-15cm --beim Laichen/spawning--
Photo: H. Linke

S72065-5 Segelflosser / Angelfish
Rotkopf, Marmor / Redhead, marbled
B, Z, 12-15cm --über Larven/over larvae--
Photo: H. Linke

S72065-5 Segelflosser / Angelfish
Rotkopf, Marmor / Redhead, marbled
B, Z, 12-15cm --mit Jungfischen/with Babies--
Photo: H. Linke

South American Cichlids IV

Skalare / Angels: Zuchtformen / Breeding forms

S72058-4 Segelflosser / Angelfish
Goldkopf, Marmor / Goldhead, marbled
B, Z, 12-15cm

Photo: J. Glaser

S72058-4 Segelflosser / Angelfish
Goldkopf, Marmor / Goldhead, marbled
B, Z, 12-15cm

Photo: J. Glaser

S72055-2 Segelflosser / Angelfish
Goldkopf, Rotwange / Goldhead, redwedge
B, Z, 12-15cm

Photo: F. Teigler

S72070-3 Segelflosser / Angelfish
Golden, spotted, „Koi-Scalar"
B, Z, 12-15cm

Photo: F. Teigler

S72070-4 Segelflosser / Angelfish
Golden, spotted, „Koi-Scalar"
B, Z, 12-15cm

Photo: H. Linke

S72070-4 Segelflosser / Angelfish
Golden, spotted, „Koi-Scalar"
B, Z, 12-15cm

Photo: H. Linke

S72070-3 Segelflosser / Angelfish
Golden, spotted, „Koi-Scalar"

Photo: F. Teigler

S72070-4 Segelflosser / Angelfish
Golden, spotted, „Koi-Scalar"
B, Z, 12-15cm

Photo: Nakano / A.C.S.

Skalare / Angels: Zuchtformen / Breeding forms

S72070-4 Segelflosser / Angelfish
Golden, spotted, „Koi-Scalar"
B, Z, 12-15cm

Photo: U. Werner

S72070-4 Segelflosser / Angelfish
Golden, spotted, „Koi-Scalar"
B, Z, 12-15cm

Photo: U. Werner

S72070-4 Segelflosser / Angelfish
Golden, spotted, „Koi-Scalar"
B, Z, 12-15cm

Photo: Nakano / A.C.S.

S72022-4 Segelflosser / Angelfish
Gold flächig / Golden, "Lutino"
B, Z, 12-15cm

Photo: Nakano / A.C.S.

South American Cichlids IV

1. S72080-5 Segelflosser / Angelfish
   Gold Perlmutt
   Golden Schildpatt
   --beim Laichen / spawning--

2. S72080-5 Segelflosser / Angelfish
   Gold Perlmutt
   Golden Schildpatt
   --mit Larven / with larvae--

Photos: H. Linke

S72025-4 Segelflosser / Angelfish
Goldkopf, Rotwange / Goldhead Redwedge
B, Z, 12-15cm
Photo: J. Glaser

S72081-4 Segelflosser / Angelfish
Silber, Goldkopf / Silver Goldhead
B, Z, 12-15cm
Photo: H.J.Mayland

S72025-3 Segelflosser / Angelfish
Goldkopf, Rotwange / Goldhead Redwedge
B, Z, 12-15cm
Photo: Nakano / A.C.S.

S72025-3 Segelflosser / Angelfish
Goldkopf, Rotwange / Goldhead Redwedge
B, Z, 12-15cm
Photo: Nakano / A.C.S.

S72080-4 Segelflosser / Angelfish
Gold Perlmutt / Golden Schildpatt
B, Z, 12-15cm
Photo: Nakano / A.C.S.

S72080-4 Segelflosser / Angelfish
Gold Perlmutt / Golden Schildpatt
B, Z, 12-15cm
Photo: H.J.Mayland

Skalare / Angels: Zuchtformen / Breeding forms

South American **Cichlids IV**

Skalare / Angels: Zuchtformen / Breeding forms

S72082-3 Segelflosser / Angelfish
Solber, Goldkopf, Schleier /
Silver Goldhead, veiltail, B, Z, 12-15cm
Photo: J. Glaser

S72021-3 Segelflosser / Angelfish
Gold, Schleier / Gold, veiltail
B, Z, 12-15cm
Photo: F. Teigler

S72090-2 Segelflosser / Angelfish
Geister Skalar / Ghost Angel
B, Z, 12-15cm
Photo: F. Teigler

S72021-2 Segelflosser / Angelfish
Gold, Schleier / Gold, veiltail
B, Z, 12-15cm
Photo: F. Teigler

S72021-2 Segelflosser / Angelfish
Gold, Schleier / Gold, veiltail
B, Z, 12-15cm
Photo: Nakano / A.C.S.

S72096-2 Segelflosser / Angelfish
Perlskalar, Albino / Pearl Angel, albino
B, Z, 12-15cm
Photo: F. Teigler

South American Cichlids IV © Verlag A.C.S. GmbH

S72020-3 Segelflosser / Angelfish
Gold, Albino / Gold, albino
B, Z, 12-15cm

Photo: F. Teigler

S72026-3 Segelflosser / Angelfish
Gold, Albino, Schleier / Gold, albino, veiltail
B, Z, 12-15cm

Photo: F. Teigler

S72090-2 Segelflosser / Angelfish
Geister-Skalar / Ghost Angel
B, Z, 12-15cm

Photo: Nakano / A.C.S.

S72090-2 Segelflosser / Angelfish
Geister-Skalar / Ghost Angel
B, Z, 12-15cm

Photo: F. Teigler

S72090-3 Segelflosser / Angelfish
Geister-Skalar / Ghost Angel
B, Z, 12-15cm

Photo: F. Teigler

S72090-2 Segelflosser / Angelfish
Geister-Skalar / Ghost Angel
B, Z, 12-15cm

Photo: F. Teigler

Skalare / Angels: Zuchtformen / Breeding forms

S72090-3 Segelflosser / Angelfish
Geister-Skalar / Ghost Angel
B, Z, 12-15cm

Photo: F. Teigler

S72091-3 Segelflosser / Angelfish
Geister-Skalar, Schwarzsaum/ Ghost Angel, blackseam
B, Z, 12-15cm

Photo: F. Teigler

S72092-2 Segelflosser / Angelfish
Geister-Skalar, Schleier / Ghost Angel, veiltail
B, Z, 12-15cm

Photo: Nakano / A.C.S.

S72093-2 Segelflosser / Angelfish
Geister-Skalar, orange / Ghost Angel, orange
B, Z, 12-15cm

Photo: F. Teigler

S72093-2 Segelflosser / Angelfish
Geister-Skalar, orange / Ghost Angel, orange
B, Z, 12-15cm

Photo: F. Teigler

S72094-2 Segelflosser / Angelfish
Geister-Skalar, Leopard / Ghost Angel, leopard
B, Z, 12-15cm

Photo: F. Teigler

South American Cichlids IV © Verlag A.C.S. GmbH

S72095-5 Segelflosser / Angelfish
Blauer Skalar, Schleier / Blue Angel, veiltail
B, Z, 12-15cm

Photo: Nakano / A.C.S.

Halten Sie Ihren AQUALOG über Jahre aktuell
Keep your AQUALOG up-to-date for years

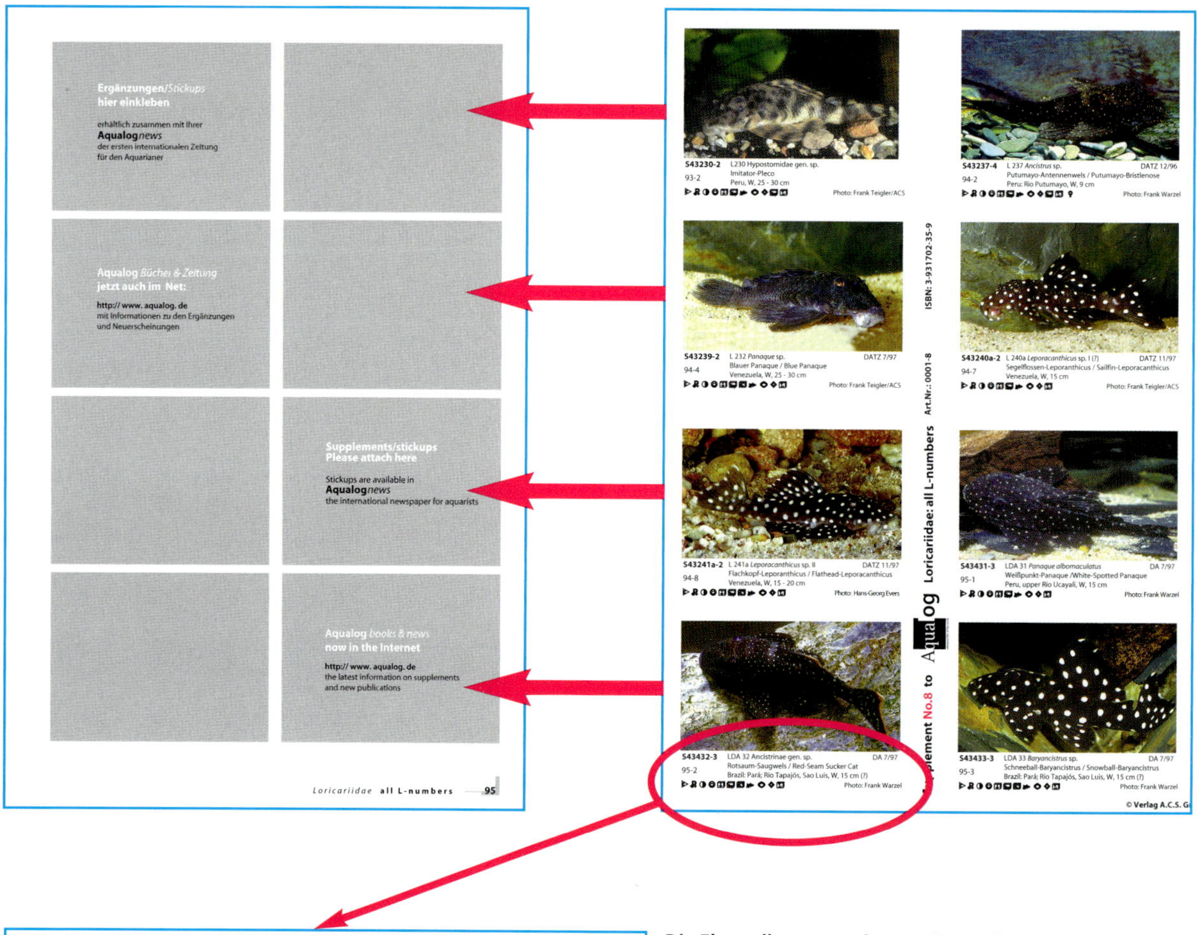

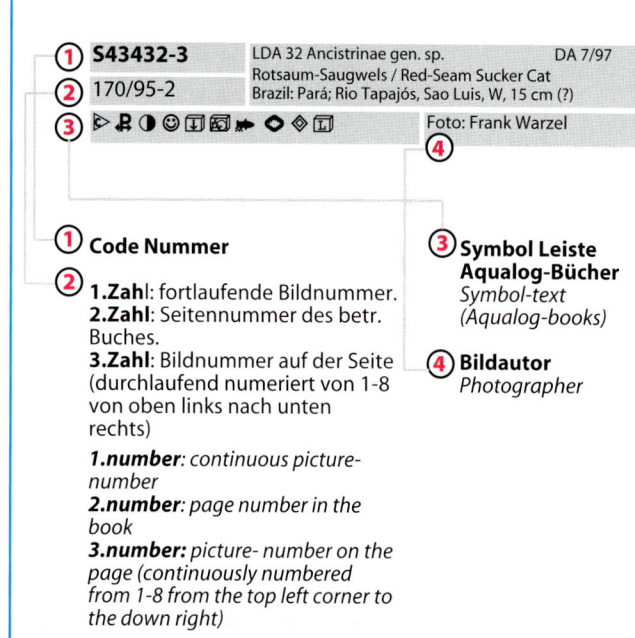

① Code Nummer

1.Zahl: fortlaufende Bildnummer.
2.Zahl: Seitennummer des betr. Buches.
3.Zahl: Bildnummer auf der Seite (durchlaufend numeriert von 1-8 von oben links nach unten rechts)

1.number: continuous picture-number
2.number: page number in the book
3.number: picture- number on the page (continuously numbered from 1-8 from the top left corner to the down right)

③ Symbol Leiste Aqualog-Bücher
Symbol-text (Aqualog-books)

④ Bildautor
Photographer

Die Flutwelle neuer oder neu-importierter Arten reißt nicht ab. Es ist leider unmöglich, sie alle in der Zeitung "AQUALOGnews" als Stickups zu präsentieren. Daher haben wir uns entschlossen, Ergänzungsbögen mit je acht Einklebebildern zu einem Buch herzustellen. Lieferbar über den guten Zoofachhandel und den Buchhandel zum Preis von DM 4.80 pro Stück. Viel Freude damit! Übrigens: die Stickups aus der news befinden sich nicht nochmals auf den Ergänzungsbögen!

The flood of new or new-imported species doesn´t stop. It is impossible to show them all as stickups in our Newspaper AQUALOGnews. So we decided to print supplements with eight stickers each (each supplement contents pictures for only one volume of AQUALOG). They can be ordered at well-equipped pet-shops or in every bookshop. We hope you enjoy them! By the way: the stickups are not reprinted on the supplements!

Bitte beachten Sie nebenstehendes Schema, bevor Sie die Bilder einkleben. Die Ergänzungen erscheinen nicht zwangsläufig in der Reihenfolge, in der sie eingeklebt werden, sondern in der Reihenfolge ihrer Verfügbarkeit. Wenn wir z.B. anfangs nur das Bild eines Weibchens als Ergänzung haben, jedoch sicher sind, früher oder später auch das Bild eines Männchens zu bekommen, sollte das Bildkästchen links vom Weibchenbild frei bleiben.

Please follow the scheme given here, before you stick in the pictures. The supplements are not necessarily in the correct order. For example: if we have only the photo of a female, but we are sure to get the photo of the male sooner or later, too, please keep the space to the left of the female free.

216   *South American Cichlids IV*   © Verlag A.C.S. GmbH

Update Service für Ihre Aqualog Bücher
Update Service for your Aqualog Books

Supplement No.1 to
Loricariidae all I-Numbers
ISBN 3-931702-15-4

Supplement No.2 to
Loricariidae all I-Numbers
ISBN 3-931702-16-2

Supplement No.3 to
Loricariidae all I-Numbers
ISBN 3-931702-17-0

Supplement No.4 to
Loricariidae all I-Numbers
ISBN 3-931702-20-0

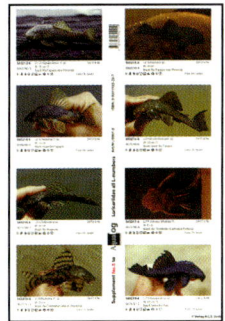
Supplement No.5 to
Loricariidae all I-Numbers
ISBN 3-931702-22-7

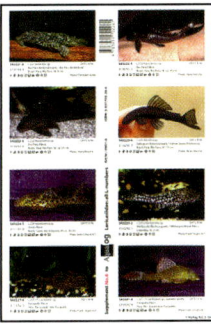
Supplement No.6 to
Loricariidae all I-Numbers
ISBN 3-931702-28-6

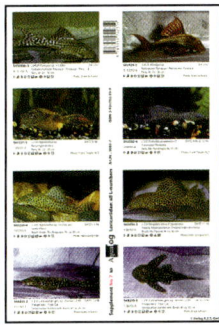
Supplement No.7 to
Loricariidae all I-Numbers
ISBN 3-931702-35-9

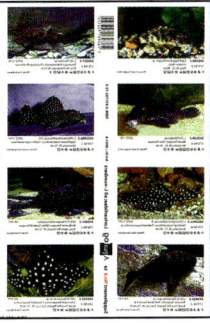
Supplement No.8 to
Loricariidae all I-Numbers
ISBN 3-931702-72-3

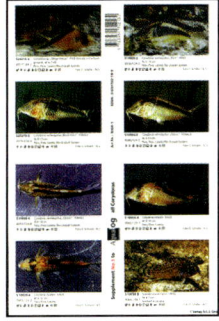
Supplement No.1 to
all Corydoras
ISBN 3-931702-18-9

Supplement No.2 to
all Corydoras
ISBN 3-931702-23-5

Supplement No.3 to
all Corydoras
ISBN 3-931702-37-5

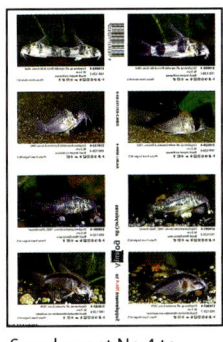
Supplement No.4 to
all Corydoras
ISBN 3-931702-83-9

Supplement No.1 to
Southamerican Cichlids 1
ISBN 3-931702-19-2

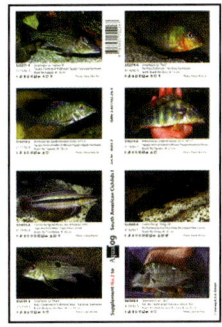
Supplement No.2 to
Southamerican Cichlids 1
ISBN 3-931702-26-X

Supplement No.1 to
Southamerican Cichlids 2
ISBN 3-931702-12-X

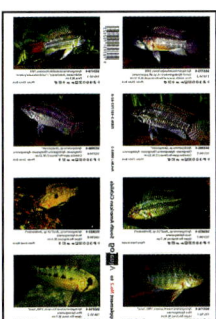
Supplement No.2 to
Southamerican Cichlids 2
ISBN 3-931702-82-0

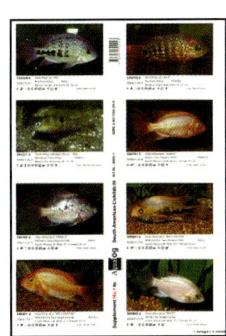
Supplement No.1 to
Southamerican Cichlids 3
ISBN 3-931702-24-3

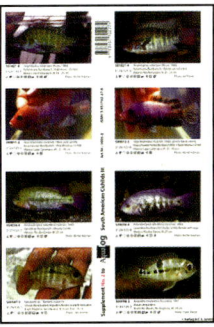
Supplement No.2 to
Southamerican Cichlids 3
ISBN 3-931702-27-8

Supplement No.1 to
all Labyrinths
ISBN 3-931702-36-7

Ergänzungen/*Stickups*
hier einkleben

erhältlich zusammen mit Ihrer
Aqualognews
der ersten internationalen Zeitung
für den Aquarianer

Supplements/stickups
Please attach here

Stickups are available in
Aqualognews
the international newspaper for
aquarists

Aqualog *Bücher & Zeitung*
jetzt auch im Net:

http:// www. aqualog. de
mit Informationen zu den Ergänzungen
und Neuerscheinungen

Aqualog *books & news*
now in the Internet

http:// www. aqualog. de
the latest information on supplements
and new publications

Ergänzungen/Stickups
hier einkleben

erhältlich zusammen mit Ihrer
Aqualognews
der ersten internationalen Zeitung
für den Aquarianer

Supplements/stickups
Please attach here

Stickups are available in
Aqualognews
the international newspaper for
aquarists

Aqualog Bücher & Zeitung
jetzt auch im Net:

http:// www. aqualog. de
mit Informationen zu den Ergänzungen
und Neuerscheinungen

Aqualog books & news
now in the Internet

http:// www. aqualog. de
the latest information on supplements
and new publications

Ergänzungen/*Stickups*
hier einkleben

erhältlich zusammen mit Ihrer
Aqualognews
der ersten internationalen Zeitung
für den Aquarianer

Supplements/stickups
Please attach here

Stickups are available in
Aqualognews
the international newspaper for
aquarists

Aqualog *Bücher & Zeitung*
jetzt auch im Net:

http:// www. aqualog. de
mit Informationen zu den Ergänzungen
und Neuerscheinungen

Aqualog *books & news*
now in the Internet

http:// www. aqualog. de
the latest information on supplements
and new publications

Ergänzungen/_Stickups_
hier einkleben

erhältlich zusammen mit Ihrer
Aqualognews
der ersten internationalen Zeitung
für den Aquarianer

Supplements/stickups
Please attach here

Stickups are available in
Aqualognews
the international newspaper for
aquarists

Aqualog _Bücher & Zeitung_
jetzt auch im Net:

http:// www. aqualog. de
mit Informationen zu den Ergänzungen
und Neuerscheinungen

Aqualog _books & news_
now in the Internet

http:// www. aqualog. de
the latest information on supplements
and new publications

Ergänzungen/*Stickups*
hier einkleben

erhältlich zusammen mit Ihrer
Aqualognews
der ersten internationalen Zeitung
für den Aquarianer

Supplements/stickups
Please attach here

Stickups are available in
Aqualognews
the international newspaper for
aquarists

Aqualog *Bücher & Zeitung*
jetzt auch im Net:

http:// www. aqualog. de
mit Informationen zu den Ergänzungen
und Neuerscheinungen

Aqualog *books & news*
now in the Internet

http:// www. aqualog. de
the latest information on supplements
and new publications

Ergänzungen/*Stickups*
hier einkleben

erhältlich zusammen mit Ihrer
Aqualognews
der ersten internationalen Zeitung
für den Aquarianer

**Supplements/stickups
Please attach here**

Stickups are available in
Aqualognews
the international newspaper for
aquarists

Aqualog *Bücher & Zeitung*
jetzt auch im Net:

http://www.aqualog.de
mit Informationen zu den Ergänzungen
und Neuerscheinungen

Aqualog *books & news*
now in the Internet

http://www.aqualog.de
the latest information on supplements
and new publications

Ihr Nachschlagewerk
your reference work!

ISBN 3-931702-04-9

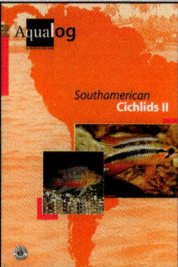
ISBN 3-931702-07-3

ISBN 3-931702-10-3

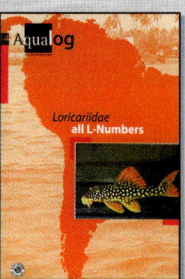
ISBN 3-931702-01-4

ISBN 3-931702-13-8

ISBN 3-931702-21-9

ISBN 3-931702-25-1

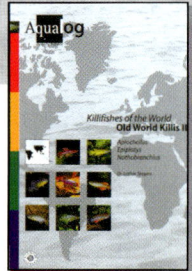
ISBN 3-931702-30-8

Demnächst *coming soon*

**Killifishes of the World
New World Killis**

Autor **Dr. Lothar Seegers**

ISBN 3-931702-77-4

Demnächst *coming soon*

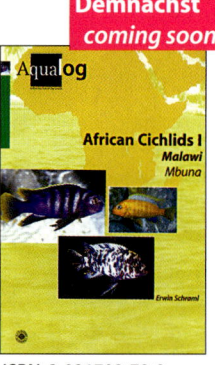
ISBN 3-931702-79-0

Demnächst *coming soon*

ISBN 3-931702-78-2

Demnächst *coming soon*

**African Cichlids I
Malawi / Mbuna**

Autor **Erwin Schraml**

**Goldfische und Schleierschwänze
all Goldfish and varieties**

Autor **Karl-Heinz Bernhardt**

Zu jedem Buch erscheinen auch dekorative Poster!
Full-colour poster to accompany each volume

Vervollständigen Sie Ihr Nachschlagewerk durch weitere Bücher der Aqualog-Reihe! *Complete this reference work with future volumes in the Aqualog series.*

Anfragen an
  For more information please contact

Verlag A.C.S. GmbH,
Liebigstraße 1, 63110 Rodgau, Fax: +49 (0)6106 - 644692, email: acs@aqualog.de

Diskus-Bewertung
von Manfred Göbel

Diskus-Ausstellungen und Wettbewerbe mit Diskusfischen werden immer umstritten sein, ebenso wie sehr viele andere Schauen und Veranstaltungen mit lebenden Tieren. Sicherlich lassen sich Argumente finden, die gegen eine solche Veranstaltung sprechen. Aber es gibt auch eine ganze Reihe von Fakten, die einen solchen Wettbewerb rechtfertigen. Dabei müssen die Bedürfnisse der ausgestellten Tiere oberste Priorität haben und die Bewertung im Wettbewerb muß so fair und objektiv wie möglich erfolgen.

Jeder Wertungsrichter muß wissen, daß er mit seiner Bewertung der ausgestellten Tiere die zukünftige Richtung in der Zucht maßgeblich beeinflußt. Daher ist Sorgfalt, Fairneß und Objektivität grundsätzliche Voraussetzung für diese Aufgabe.

Bei Diskus-Wettbewerben besteht die Hauptproblematik darin, etwas bewerten zu müssen, das sich rational nicht messen läßt. Das ist zwar ein Riesenproblem, aber es ist nicht neu. Auf anderen Gebieten findet man das gleiche Problem ebenfalls, so z.B. in der Kunst, der Meinungsforschung, der Werbung beim Eiskunstlaufen oder dem Turmspringen.

Darüber hinaus stellen sich auch noch andere Probleme:

1. Für die Bewertung wirklich relevante Kriterien zu finden.
2. Die unterschiedlichen Farbvarianten zu berücksichtigen.
3. Die Preisrichter nicht durch eine an sich notwendige Unmenge von Einzelinformationen zu überfordern.
4. Das Problem der psychischen Überlastung der Preisrichter durch die Notwendigkeit, unzählige Male emotionale Eindrücke in kürzester Zeit in Rationalität = Punkte umsetzen zu müssen.
5. Das Problem, daß jeder Preisrichter, bewußt oder unbewußt, persönliche Vorlieben und Vorurteile hat.

Wir haben dadurch drei Problembereiche:
- den rein Diskus-bezogenen Bereich,
- den physischen Bereich (bezogen auf die Anstrengung der Preisrichter),
- den psychischen (ebenfalls bezogen auf die Anforderung an die Preisrichter)

Wenn wir zu einer einigermaßen gerechten Lösung kommen wollen, müssen wir alle drei Problemkreise zufriedenstellend lösen. Lassen Sie uns mit dem „Diskus-bezogenen" Bereich beginnen. Alle Diskus-Varianten werden üblicherweise einer von neun Gruppen zugeordnet. Da diese Einteilung weltweit Verwendung findet und weitestgehend aner-

Assessment of Discus
by Manfred Göbel

Discus exhibitions and competitions will always be disputed, as well as many other shows and events dealing with alive animals. There will surely be found reasons against such an event. However, there are lot of facts justifying such a competition. Priority must be given to the exhibited animals´ needs and the assessment of the contest has to be as fair and objective as possible.

Every judge has to know that by evaluating the exhibited animals he has a considerable influence on the future direction of breeding. Therefore, carefulness, fairness and objectivity are absolutely necessary conditions for this task. The main problem of discus competitions lies in having to assess something, that cannot be measured rationally. In fact, this is a huge problem, but it is nothing new. In other fields you can find the same problem as well, e.g. in art, opinion research, advertising, ice-skating or in high diving.

Moreover, the following problems also occur:

1. To find really relevant criteria for the assessment.
2. To take the different variants in colour into consideration.
3. Not to demand too much of the judges by loads of single criteria, which are actually necessary.
4. Psychological strain of the judges due to the necessity to transform emotional impressions innumerable times into rationality = points; and all this in no time.
5. The problem that every judge has his personal likings as well as prejudices - consciously or unconsciously.

Therefore, there are three problem fields:

- *the one dealing with discus only*
- *the physical one and*
- *the psychological field*

If we want to come to a quite fair conclusion, we have to solve all of the three problem fields in a satisfactory way. Let us start with the field concerning discus.
All discus variants normally belong to one of nine groups. As this classification is applied world-wide and accepted to a great extend, the assessment is founded on it.

kannt ist, bildet sie die Basis der Bewertung.

Die neun Gruppen sind:

A = Heckel-Diskus
    Wildfänge und reine Nachzuchten
B = Brauner Diskus
    Wildfänge und reine Nachzuchten
C = Blauer Diskus
    Wildfänge und reine Nachzuchten
D = Grüner Diskus
    Wildfänge und reine Nachzuchten
E = Diskus türkis gezeichnet
F = Diskus flächig-türkis
G = Diskus rot-türkis
H = Diskus flächig rot
I = Diskus offene Klasse
    (Sonderformen wie Pidgeon Blood etc.)

Folgende Bewertungskriterien haben sich in der Diskussion herauskristallisiert:

1 - Gesamteindruck
    (international: „Overall impression")
2 - Körpergröße
    (international: „Body size")
3 - Proportionen und Harmonie
    (international: „Proportions and harmony")
4 - Augen
    (international: „Eyes")
5 - Flossen und Kiemen
    (international: „Fins and gills")
6 - Körperzeichnung
    (international: „Body markings")
7 - Farbe: Blau- und Grüntöne
    (international: „Colour: blue and green shades")
8 - Farbe: Braun-, Gelb- und Rottöne
    (international: „Colour: Brown, yellow and red shades")

Wenn man diese Punkte nun anschaut, wird man feststellen, daß dies möglicherweise die wichtigsten Punkte sind, daß sie aber letztendlich trotzdem nicht ausreichen, um z.B. den unterschiedlichen Farbschlägen gleichermaßen gerecht zu werden. So ist z.B. bei einem Grünen Diskus eine runde Körperform und damit Proportion und Harmonie sicherlich höher zu bewerten, als bei einem Heckel-Diskus, bei dem ein harmonischer Körperbau fast selbstverständlich ist. Zeigt ein Heckel-Diskus eine intensive blaue Schmuckfarbe, so ist dies höher zu bewerten, als bei einem Blauen Diskus. Dieser Aspekt der unterschiedlichen Farbvarianten mit ihren unterschiedlichen Wertigkeiten bei glei-

The nine categories are as follows:

A = Heckel-Discus
    wild caught and purely bred discus
B = Brown discus
    wild caught and purely bred discus
C = Blue discus
    wild caught and purely bred discus
D = Green discus
wild caught and purely bred discus
E = Striped turquoise
F = Turquoise discus
G = Red Turquoise
H = Red discus
I = Open class
    (Unclassifiable forms)

The following assessment criteria have crystallised in discus breeding:

1. Overall impression
2. Body size
3. Proportions and harmony
4. Eyes
5. Fins and gills
6. Body markings
7. Colour: blue and green shades
8. Colour: brown yellow and red shades

If you now look at these points, you will notice that these are possibly the most important ones. However, finally they will not be sufficient anyway, e.g. to do justice to the different categories in the same way. Hence, the round body and therefore proportions and harmony of a Green Discus would have to be given more points than a Heckel-Discus with the same shape, because for the latter a harmonious body is quite natural. If a Heckel-Discus shows an intensive blue colour, it has to be given more points than a Blue Discus.

This aspects of the different variants in colour with its different valencies has to be considered at any rate when evaluating the same criterion, because only this way a fair assessment is possible.

Every colour variant has its different main criteria, which has to be thoroughly considered. This could happen in three ways:

chen Kriterien muß auf jeden Fall berücksichtigt werden, da nur so eine faire Bewertung möglich ist.

Jede Farbvariante hat unterschiedliche Schwerpunkte, auf die eingegangen werden muß. Das könnte auf drei Arten geschehen:

1 - man könnte neue, zusätzliche Gruppen schaffen. Aber wenn man dabei alle Farbvarianten mit allen vorstellbaren Körperzeichnungen berücksichtigen wollte, käme man leicht auf über 100 Wertungsklassen. das mag zwar gerechter sein, ist aber undurchführbar.
2 - Man könnte die Zahl der Wertungskriterien erhöhen. Damit überfordert man aber die Preisrichter, vor allem, wenn es gilt, hunderte von Fischen zu bewerten.
3 - Durch gezielte Vorarbeit bedient man sich bestimmter Mechanismen, die zu einer Differenzierung führen, ohne daß es zu einer Mehrbelastung der Preisrichter kommt.

Genau danach habe ich gesucht und mich dabei Mechanismen bedient, die es in anderen, vergleichbaren Bereichen schon gibt. Die Lösung des Problems besteht meines Erachtens darin, den jeweiligen Gruppen im Verhältnis zu den einzelnen Kriterien sogenannte „Handicap-Faktoren" zuzuordnen und zwar entsprechend der Bedeutung der einzelnen Punkte in Bezug auf die jeweilige Farbvariante. Man baut einen „Multiplikator" ein, der die Wertigkeit des jeweiligen Kriteriums in Beziehung zu der jeweiligen Variante setzt.. Das bedeutet, daß z.B. beim Punkt „Proportionen und Harmonie" der Multiplikator bei einem Grünen Diskus größer ist als bei einem Heckel-Diskus, der des Punktes „Körpergröße" aber erheblich kleiner ist als z.B. bei der „Offenen Klasse". Dadurch werden die einzelnen Klassen erst miteinander vergleichbar. Voraussetzung ist allerdings, daß die Summe der einzelnen Multiplikatoren für alle Varianten die gleiche Zahl ergibt.

Das klingt alles sehr theoretisch und kompliziert, ist in Wirklichkeit aber ganz einfach, wenn die Vorarbeiten gründlich durchdacht worden sind und wenn man die Verarbeitung dann durch den Computer erledigen läßt. Für den Preisrichter selbst wird die Sache in keiner Weise schwieriger oder umfangreicher. Er hat lediglich neben seinen acht Einzelbewertungen nur noch seine persönliche Preisrichter-Nummer auf dem Bewertungsbogen anzukreuzen. Alles andere wird dann vom Computer erledigt. Auf diese Weise, durch die Einschaltung eines „Handicap-Faktors", erreichen wir ohne Mehrarbeit für den Preisrichter eine größtmögliche Differenzierung.

Diese Methode hat den Vorteil, daß sich der Preisrichter

1. We could create new additional groups. But if we wanted to take all colour variants with every imaginable pattern into consideration, we could easily exceed 100 assessment categories. That may be fairer, but it is impractical.
2. We could increase the number of assessment criteria. This way we would demand too much of the judges though, especially when it comes to the evaluation of hundreds of discus.
3. By specific preparations we make use of certain mechanisms which lead to a distinction without causing additional strain for the judges.

This is exactly what I have been looking for and I used those mechanisms which already exist in other comparable fields. In my view the solution is to attach so-called „handicap-factors" to the respective groups in proportion to the individual criteria, namely corresponding to the importance of each point with reference to the respective colour variant. You add a „multiplier" which relates the value of each criteria to the respective variant. This means, for example, that for the point „proportions and harmony" the multiplier is bigger with a Green Discus than with a Heckel Discus, for the point „body size", however, is considerably smaller for a green one than for a discus of an unclassifiable form.

This way the different colour variants only become comparable though. However, only on condition that the sum of the individual multipliers is the same for all variants.
This sounds all very theoretical and complicated, but in fact it is quite simple, if the preparations have been well thought-out and if it is then processed by computer. For the judge himself the matter does not become more difficult or extensive at all. Besides his eight single assessments he only has to tick off his personal judge number on the evaluation sheet. Anything else is done by the computer then. This way, by adding a „handicap-factor", we achieve the biggest possible differentiation without any additional work for the judge.

This method has the advantage that the judge can totally concentrate on each assessment criteria in detail without having to think of any other aspects. As the judges are not aware of the actual „handicap-factors" on the assessment sheet, moreover, the danger of conscious or unconscious manipulation is reduced considerably. If you now act according to this procedure, an assessment sheet looks

ganz auf die jeweiligen Bewertungskriterien im Detail konzentrieren kann, ohne ständig an übergeordnete Aspekte denken zu müssen. Da den Preisrichtern die tatsächlichen „Handicap-Faktoren" auf den Bewertungsbögen nicht gegenwärtig sind, wird die Gefahr der bewußten oder unbewußten Manipulation außerdem noch erheblich verringert. Wenn man nun nach diesem Verfahren vorgeht, sieht ein Bewertungsbogen trotz der zahlreichen Differenzierung, die in ihm steckt sehr einfach aus. Die Einordnung des zu bewertenden Fisches in eine der neun Gruppen, die ausschließlich durch den Besitzer erfolgt, und die Bewertung nach den acht vorgegebenen Kriterien setzt unter Berücksichtigung der „Handicap-Faktoren" eine Reihe von Rechenschritten in Gang, die eine größtmögliche Differenzierung bei der Bewertung ergeben. Soviel zu den Lösungsmöglichkeiten für den Diskus-bezogenen Problembereich und einen Teil der physischen Problematik.

Was übrig bleibt, ist die psychische Problematik, die es bei jedem Bewertungsverfahren gibt. Sie liegt darin, daß sich kein Mensch freimachen kann von Vorlieben und Vorurteilen. Ein Teil dessen ist durch den „Handicap-Faktor" relativiert worden. Das größte psychische Problem liegt jedoch in der Notwendigkeit, emotionale Eindrücke hunderte von Malen in kürzester Zeit in Rationalität, in Zahlen, umsetzen zu müssen. Das müssen wir zwar in vielen Bereichen tagtäglich und wird von uns als selbstverständlich empfunden, ist tatsächlich aber ungeheuer schwer und ermüdend. Mag sein, daß es dem einen leichter fällt, als dem anderen. Wie schwierig es letztlich doch ist, können wir alle an uns selbst feststellen. Wir alle werden mit ziemlicher Sicherheit bei einer Bewertung den 37. Fisch weniger sorgfältig und aufmerksam bewerten, als die ersten zehn. Das liegt ausschließlich an dem Ermüdungsprozeß, der durch die ständigen Rationalisierungsversuche hervorgerufen wird.

Hierbei muß dem Preisrichter geholfen werden und auch dazu können wir den Computer einsetzen. Ihm möchten wir diesen ständig wiederkehrenden Rationalisierungsprozeß überlassen. Dem Preisrichter dagegen sollte die Möglichkeit geboten werden, seine emotionalen Eindrücke auch emotional artikulieren zu können, indem man ihm statt Punktzahlen emotionale Formulierungen zur Bewertung anbietet. Hierzu eignen sich die allseits bekannten Schulnoten:

sehr gut (international: „very good")
gut (international: „good")
befriedigend (international: „fair")
mangelhaft (international: „poor")

very simple, in spite of the comprehensive differentiation that it contains.

Considering the „handicap-factors", the classification of the fish to one of the nine groups which is only done by the proprietor, and the evaluation according to the eight given criteria, get a couple of calculating steps going which result in the best possible differentiation when assessing the fish. So is it for the solution possibilities for the problem field concerning discus as well as for a part of the physical problems.

What remains is the psychological problem existing in every test procedure, namely that nobody can claim to be completely free of liking and prejudices. Part of it has been qualified by the „handicap-factor". However, the biggest psychological problem is the necessity to have to turn emotional impressions a several hundred times into rationally and numbers - and that in no time. In fact, this is what happens to many of us in many fields day by day and we take it for granted. Indeed, it is quite difficult and tiring though. It may be that one finds it easier than another. But everybody can observe how difficult it is at the end of the day. With a fair degree of certainty we all will assess the fish number 37 with less thoroughness and attention than the first ten fishes. This is due to the process of tiredness caused by the constant rationalisation attempts.

In this connection the judge needs help and this help can come from a computer. It is the computer we would like to leave this constant rationalisation process to. The judge, however, should have the possibility to emotionally articulate his emotional impressions by offering him emotional formulations instead of points for the assessment. For this purpose the well known school notes can be used:

- *very good*
- *good*
- *fair*
- *poor*

If a discus exceeds all his competitors in one category he can also be awarded with the note **„excellent"**: These notes will then be transformed by the computer into matter-of-fact figures and result in a total number of points together with the above mentioned „handicap-factors" which determines the finishing order.

Ist ein Diskus in einem Bereich herausragend besser als alle vergleichbaren, kann auch die Note **„vorzüglich"** (international: „excellent") vergeben werden. Diese Noten wiederum werden später vom Computer in nüchterne, der Wertigkeit entsprechende Zahlenwerte umgesetzt und ergeben dann zusammen mit den oben genannten „Handicap-Faktoren" eine Gesamtpunktzahl, die ihrerseits die Plazierung ergibt. Dieses Verfahren ist eine echte Erleichterung für den Preisrichter und ist in ähnlicher Form im Bereich der Produkt-Testverfahren, der Meinungsforschung, ja eigentlich bei jeder Analyse im mehr oder weniger irrationalen Bereich heute normal. Unter Einbeziehung all dieser Aspekte und unter Berücksichtigung der Tatsache, daß im Computer unter der laufenden Nummer der Diskus in der Gruppe der jeweiligen Farbvariante bereits gespeichert ist, ergibt sich die Form des Bewertungsbogens. Er ist einfach und enthält dennoch die größtmögliche Differenzierung, sowie eine weitgehende Nivellierung menschlicher Unzulänglichkeiten und dürfte damit die größtmögliche „Gerechtigkeit" sicherstellen (soweit das in einem irrationalen und emotionalen Bereich überhaupt möglich ist).

Weitere Vorteile sind:

1. Durch das von einem Spezialisten geschriebene, vorliegende Programm ist die Eingabe schnell und unkompliziert, wobei eine Anzahl von Preisrichtern von 1 - 5 möglich ist.
2. Bedienungsfehler sind nahezu ausgeschlossen und man kann in Sekunden das gewünschte Einzelergebnis erhalten.
3. Eine Gesamtübersicht (z.B. welcher Preisrichter welchen Fisch noch nicht bewertet hat) ist jederzeit abrufbar.
4. Jeder Fisch kann unter seiner laufenden Nummer abgerufen und überprüft werden.
5. Eine auf Farbvarianten bezogene Auswertung und Rangfolge auf der Basis der bisher erfolgten Eingaben ist zu jedem Zeitpunkt möglich.
6. Für alle Fische jeder Gruppe können Listen entsprechend der Reihenfolge der erteilten Wertungspunkte ausgedruckt werden.

Das vorliegende computergestützte System kann die Arbeit bei der Bewertung von Diskusfischen wesentlich erleichtern. Die Verantwortung für eine faire, objektive Bewertung von Diskusfischen liegt allerdings nach wie vor in den Händen der jeweiligen Wertungsrichter. Das Fachwissen der Bewerter sowie ihre Sorgfalt und Objektivität bei der Beurteilung der einzelnen Fische sind die Grundvoraussetzungen für eine bestmögliche Diskus-Bewertung.

This procedure is a real facilitation for the judge and in a resembling way today it is often used in the more or less irrational area of product testing, market research and actually every analysis. Including all these aspects and taking into account that the computer has already stored the discus under the serial number in the group of the relevant colour variant the form of the assessment sheet can be derived. It is simple but nevertheless it contains the greatest possible differentiation as well as an extensive levelling of human insufficiency and it will probably secure the greatest possible fairness (as far as it is possible in such an irrational and emotional field).

Further advantages:

***1.** The programme on hand which is written by an expert qualifies for an easy and uncomplicated input. A number of 1 to 5 judges is possible.*
***2.** Operating mistakes are nearly impossible and within seconds you can get the requested individual result.*
***3.** A total survey (e.g. which judge has not assessed which fish) can be recalled at any time.*
***4.** Each fish can be recalled and checked under his serial number.*
***5.** An assessment regarding the colour variants and the order of preference can be done at any time on the base of the data which have already been stored.*
***6.** It is possible to print lists for all fishes of each group according to the order of the given assessment points.*

This computer based system can significantly ease the judge's work of assessing discusfish. However, the responsibility for a fair and objective assessment is still in hands of the respective judges. Their special know-how as well as their carefulness and objectivity when assessing the fish are the basic conditions for a best possible assessment of discus.

Für den Bewerter:

Bewerten Sie den Diskus bitte so, wie er sich zum Zeitpunkt der Bewertung im Becken zeigt. Versuchen Sie bitte **nicht**, vorangegangenen Transport, falsche Beleuchtung und ähnliche Streßfaktoren mit zu berücksichtigen.

Versuchen Sie bitte **nicht**, den Diskus im Vergleich zu anderen Diskusfischen zu bewerten.

Lassen Sie einfach den zu bewertenden Fisch auf sich einwirken und markieren Sie dann Ihre Eindrücke auf dem Bewertungsbogen.

Alle weiteren zur Bewertung wichtigen Faktoren (Farbschlag usw.) werden im Computerprogramm automatisch berücksichtigt.

Tragen Sie bitte im Kopf des Bogens in die jeweilige Spalte die Ident-Nummer des zu bewertenden Fisches und Ihre eigene Bewerter-Nummer ein. Bewerten Sie den Diskus nun, indem Sie bei jedem der acht Wertungspunkte den Ihrer Meinung nach zutreffenden Wert ankreuzen.

Werte:

excellent: Sollte nur vergeben werden, wenn der Fisch in dem zu wertenden Punkt außergewöhnliche Vorzüge aufweist.
very good: Wenn der Fisch in dem entsprechenden Punkt zur Spitzenklasse zählt.
good: Im zutreffenden Punkt über dem Durchschnitt.
fair: Durchschnitt
poor: Deutlich unter dem Durchschnitt.

Wertungspunkte:

1. Gesamteindruck:
Wirkung auf den Betrachter; Ausdrucksstärke des Fisches; Gesundheit und Wohlbefinden; Verhalten im Aquarium.
2. Körpergröße:
Gesamtgröße und Mächtigkeit des Fisches
3. Proportionen und Harmonie:
Maßverhältnisse: Länge zu Höhe zu Breite; Übergänge: Maul zu Stirn (Stirnhöcker?); Stirn zu Rückenflosse (Absatz?); Maul zu Kehle und Brust; Brust zu Afterflosse.
4. Flossen und Kiemen:
Flossen: einwandfrei feine Ausbildung? Grobe Flossenstrahlen oder Deformationen?
Kiemen: Einwandfreie Ausbildung der Kiemendeckel? Kiemendeckel abstehend oder aufgerollt? Deformationen?

Information for the judge:

Please evaluate the discus according to its appearance in the aquarium at the time of the evaluation. Please **do not** attempt to take the preceding transport, incorrect lighting or similar stress factors into consideration.

Please **do not** attempt to evaluate this discus in comparison to other discus fish.

Simply look at the fish to be evaluated and mark your impression on the evaluation sheet. All other factors significant for the evaluation will be automatically taken into consideration by the computer programme.

Please enter the Id. No. of the fish to be evaluated and your own judge No. at the top of each sheet. Now evaluate the discus by marking a cross next to the value of each of the eight evaluation points which is appropriate in your opinion.

Values:

excellent: Should only be marked if the fish excels in the point to be evaluated.
very good: When the fish ranks among the top of its class in the respective point.
good: Above average in the point to evaluated.
fair: Average.
poor: Clearly below average.

Evaluation points:

1. Overall impressions:
Effect on the observer. Expressiveness of the fish. Health and well-being; Behaviour in the aquarium.
2. Body size:
Overall size and massiveness of the fish.
3. Proportions and harmony:
Dimensional relationships: Length to height to width; Muscular system to fins.
Transitions: Mouth to forehead (bulging forehead?); Forehead to dorsal fin (indention?); Mouth to throat and chest; Chest to anal fin.
4. Fins, tail and gills:
Fins and tail: Perfect, fine development? Coarse fin rays or deformations?
Gills: Perfect development of the gill covers? Gill covers sticking out or rolled in? Deformations?

5. Eyes:

Eye colour and brightness; Dimensional relationship of eye size to body size.

6. Body markings:

Beauty and harmony of the markings; Uniformity throughout all body parts; Overall surface coloration of „single colour" fish; Colour delimitation in striped fish.

7. Colour blue and green shades:

Clarity of the colour (is the colour pure or faded?); Brightness (intensive or pale?).

8. Colour yellow and red shades:

Clarity of the colour (brown-red or red?); Red with a tinge of blue? Tomato red or orange? Brightness (intensive or pale?).

Once the evaluation is completed, the evaluation data is entered into the computer, converted into points, multiplied by the applicable „multiplication factor" according to the colour of the fish, and the overall points are determined.

These overall points are the divided by the number of participating judges to obtain the average point value. Every fish can reach a maximum of 100%, the average point value is then divided by a factor to express the points as a percentage.

As an example, the evaluation may read as follows: Discus No. 22 Final Result = 78,7%

One request When acting as a judge, please make every effort to be **objective** and **fair** because it is quite possible that the discus you are presently evaluating first saw the light of day in your own breeding aquaria.

INDEX
Code - Nummern / Code - numbers

Code	Genus	Species	Description	Page
S71900	Pterophyllum	altum	Hoher Segelflosser/Real Altum-Angel, wild	181/82
S71901	Pterophyllum	altum	Hoher Segelflosser/Real Altum-Angel, wild, Rio-Negro	182
S71905	Pterophyllum	altum	Hoher Segelflosser/Real Altum-Angel, F1-NZ/bred	182
S71907	Pterophyllum	altum	Hoher Segelflosser/Real Altum-Angel, F2-NZ/bred	182
S71910	Pterophyllum	leopoldi cf.	Leopold's Skalar/Angel, WF-NZ, bred	184
S71912	Pterophyllum	leopoldi	Leopold's Skalar/Angel, Peru, wild	184
S71916	Pterophyllum	sp. ("dumerilii")	Schafskopf-Skalar/Sheepshead-Angel, wild	190
S71917	Pterophyllum	sp. ("dumerilii")	Schafskopf-Skalar/Sheepshead-Angel, wild	191
S71918	Pterophyllum	sp. ("dumerilii")	Schafskopf-Skalar/Sheepshead-Angel, wild	191
S71919	Pterophyllum	sp. ("dumerilii")	Schafskopf-Skalar/Sheepshead-Angel, wild	191
S71920	Pterophyllum	sp. ("dumerilii")	Schafskopf-Skalar/Sheepshead-Angel, wild	191
S71921	Pterophyllum	sp. ("dumerilii")	Schafskopf-Skalar/Sheepshead-Angel, wild	191
S71922	Pterophyllum	sp. ("dumerilii")	Schafskopf-Skalar/Sheepshead-Angel, wild	191
S71925	Pterophyllum	scalare cf.	Peru-Altum, wild	181
S71926	Pterophyllum	scalare cf.	Peru-Altum, wild	189
S71927	Pterophyllum	scalare cf.	Peru-Altum, wild	189
S71928	Pterophyllum	scalare cf.	Peru-Altum, spotted, wild	189
S71930	Pterophyllum	scalare cf.	Peru-Altum, "Variante II", wild	190
S71932	Pterophyllum	sp. (two species?)	Rio Negro-Skalar/Angel (leg. den Daas)	190
S71933	Pterophyllum	sp. (two species?)	Rio Arian-Skalar/Angel (leg. den Daas)	190
S71934	Pterophyllum	sp. (two species?)	Rio Cuiuni-Skalar/Angel (leg. Dawes)	190
S71935	Pterophyllum	scalare	Skalar/Angel, wild	186
S71936	Pterophyllum	scalare	Skalar/Angel, wild	186
S71937	Pterophyllum	scalare	Skalar/Angel, wild	186
S71939	Pterophyllum	scalare	Skalar/Angel, (sogen./so-called "eimekei"), Brazil, wild	187
S71940	Pterophyllum	scalare	Skalar/Angel, (sogen./so-called "eimekei"), NZ/bred	187
S71945	Pterophyllum	scalare	Skalar/Angel, "Red-Spotted", wild GUYANA	188
S71947	Pterophyllum	scalare	Skalar/Angel, "Red-Spotted", wild PERU	188/89
S72000	Pterophyllum	scalare	Segelflosser/Angelfish "Vierstreifen/Four-Stripes"	194
S72005	Pterophyllum	scalare	Segelflosser/Angelfish Hell, Normalflosser/Bright, normal finned	195/96
S72006	Pterophyllum	scalare	Segelflosser/Angelfish Hell, Schwarzflosser/Bright, black finned	195
S72007	Pterophyllum	scalare	Segelflosser/Angelfish Zebra, hell/Zebra, bright	199
S72008	Pterophyllum	scalare	Segelflosser/Angelfish Zebra, hell, Hochflosser/Zebra, bright, hifin	199
S72010	Pterophyllum	scalare	Segelflosser/Angelfish Hell, Schleier/Bright, veiltail	195/99
S72011	Pterophyllum	scalare	Segelflosser/Angelfish Hell Schleier /Bright, veiltail-hifin	196
S72012	Pterophyllum	scalare	Segelflosser/Angelfish Hell Hochflosser/Bright hifin	196/99
S72013	Pterophyllum	scalare	Segelflosser/Angelfish Rauchskalar, Normalflosser/Black Lace Angel	197
S72014	Pterophyllum	scalare	Segelflosser/Angelfish Rauchskalar, Breitflosser/Black Lace Angel broadfin	197
S72015	Pterophyllum	scalare	Segelflosser/Angelfish Schwarzstreifen/Black stripe	198
S72016	Pterophyllum	scalare	Segelflosser/Angelfish Schwarzstreifen, Schleier/Black stripe, veiltail	198
S72020	Pterophyllum	scalare	Segelflosser/Angelfish Gold, Albino/Gold, albino	213
S72021	Pterophyllum	scalare	Segelflosser/Angelfish Gold Schleier/Golden, veiltail	212
S72022	Pterophyllum	scalare	Segelflosser/Angelfish Gold flächig/Golden "Lutino"	209
S72025	Pterophyllum	scalare	Segelflosser/Angelfish Goldkopf, Rotwange/Goldhead, Redwedge	211
S72026	Pterophyllum	scalare	Segelflosser/Angelfish Gold, Albino, Schleier/Gold, albino, veiltail	213
S72030	Pterophyllum	scalare	Segelflosser/Angelfish Zebra Normalflosser/Zebra, normal finned	200
S72031	Pterophyllum	scalare	Segelflosser/Angelfish Rauchskalar, Zebra/Black Lace Angel, zebra	200
S72032	Pterophyllum	scalare	Segelflosser/Angelfish Rauchskalar, Zebra, Schleier/Black Lace Angel, zebra, veiltail	200
S72035	Pterophyllum	scalare	Segelflosser/Angelfish Leopard Normalfosser/Leopard, normal finned	200/06
S72036	Pterophyllum	scalare	Segelflosser/Angelfish Leopard, Schleier/Leopard, veiltail	206
S72037	Pterophyllum	scalare	Segelflosser/Angelfish Leopard, Goldkopf, Schleier./Leopard, Goldhead, veiltail	206
S72038	Pterophyllum	scalare	Segelflosser/Angelfish Leopard, Goldkopf/Leopard, Goldhead	206
S72040	Pterophyllum	scalare	Segelflosser/Angelfish Halbschwarz/Black-White "Bicolor"	201
S72041	Pterophyllum	scalare	Segelflosser/Angelfish Halbschwarz, Marmor/Black-White marbled	201
S72042	Pterophyllum	scalare	Segelflosser/Angelfish Halbschwarz, Gold/Black-White, gold	201
S72050	Pterophyllum	scalare	Segelflosser/Angelfish Halbschwarz Rotwange/Black-White, redwedge	201
S72051	Pterophyllum	scalare	Segelflosser/Angelfish Gold, Marmor/Golden, marbled	203/4
S72052	Pterophyllum	scalare	Segelflosser/Angelfish Goldkopf, Marmor/Goldhead, marble "FALB-SCALAR"	203
S72053	Pterophyllum	scalare	Segelflosser/Angelfish Gold-Marmor, Schleier/Gold-Marble, veiltail	204
S72054	Pterophyllum	scalare	Segelflosser/Angelfish Gold-Marmor Hochflosser/Gold-Marble hifin	204
S72055	Pterophyllum	scalare	Segelflosser/Angelfish Goldkopf, Rotwange/Goldhead, redwedge	208
S72056	Pterophyllum	scalare	Segelflosser/Angelfish Goldkopf, Schleier/Goldhead, veiltail	205
S72057	Pterophyllum	scalare	Segelflosser/Angelfish Schwarz, Gold-Marmor, Schleier./Black, Gold-Marble, veiltail	205
S72058	Pterophyllum	scalare	Segelflosser/Angelfish Goldkopf, Marmor/Goldhead, marbled	208
S72060	Pterophyllum	scalare	Segelflosser/Angelfish Schwarz, Normalflosser/Black, normal finned	202
S72061	Pterophyllum	scalare	Segelflosser/Angelfish Schwarz, Breitflosser/Black, broadfin	202
S72062	Pterophyllum	scalare	Segelflosser/Angelfish Schwarz, Schleier/Black, veiltail	202
S72063	Pterophyllum	scalare	Segelflosser/Angelfish Schwarz, Gold-Marmor/Black, Gold-Marble	203
S72064	Pterophyllum	scalare	Segelflosser/Angelfish Schwarz, Gold-Marmor, Schleier/Black, Gold-Marble, veiltail	203
S72065	Pterophyllum	scalare	Segelflosser/Angelfish Rotkopf, Marmor/Redhead, marbled	207
S72070	Pterophyllum	scalare	Segelflosser/Angelfish Gold, spotted "KOI-Scalar"	208
S72080	Pterophyllum	scalare	Segelflosser/Angelfish Gold Perlmutt/Golden Schildpatt	210/11
S72081	Pterophyllum	scalare	Segelflosser/Angelfish Silber-Goldkopf/Silver-Goldhead	211
S72082	Pterophyllum	scalare	Segelflosser/Angelfish Silber-Goldkopf Schleier/Silver-Goldhead, veiltail	212
S72090	Pterophyllum	scalare	Segelflosser/Angelfish Geister-Scalar/Ghost-Angel	212-14
S72091	Pterophyllum	scalare	Segelflosser/Angelfish Geister-Skalar Schwarzsaum/Ghost-Angel, blackseam	214
S72092	Pterophyllum	scalare	Segelflosser/Angelfish Geister-Skalar Schleier/Ghost-Angel, veiltail	214
S72093	Pterophyllum	scalare	Segelflosser/Angelfish Geister-Skalar, orange/Ghost Angel, orange	214
S72094	Pterophyllum	scalare	Segelflosser/Angelfish Geister-Skalar, Leopard/Ghost Angel, leopard	214
S72095	Pterophyllum	scalare	Segelflosser/Angelfish Blauer Skalar , Schleier/Blue Angel, veiltail	215
S72096	Pterophyllum	scalare	Segelflosser/Angelfish Perlskalar, Albino/Pearl Angel, albino	212
S73101	Symphysodon	aequifasciatus	DISCUS "TÜRKIS/TURQUOISE"	105
S73102	Symphysodon	aequifasciatus	DISCUS "TÜRKIS/TURQUOISE"	105
S73103	Symphysodon	aequifasciatus	DISCUS "TÜRKIS/TURQUOISE"	105
S73104	Symphysodon	aequifasciatus	DISCUS "TÜRKIS/TURQUOISE"	105
S73105	Symphysodon	aequifasciatus	DISCUS "TÜRKIS/TURQUOISE"	105
S73106	Symphysodon	aequifasciatus	DISCUS "TÜRKIS/TURQUOISE"	105
S73107	Symphysodon	aequifasciatus	DISCUS "TÜRKIS/TURQUOISE"	106
S73108	Symphysodon	aequifasciatus	DISCUS "TÜRKIS/TURQUOISE"	106
S73109	Symphysodon	aequifasciatus	DISCUS "TÜRKIS/TURQUOISE"	106
S73110	Symphysodon	aequifasciatus	DISCUS "TÜRKIS/TURQUOISE"	106
S73111	Symphysodon	aequifasciatus	DISCUS "TÜRKIS/TURQUOISE" "COBALT"	106
S73112	Symphysodon	aequifasciatus	DISCUS "TÜRKIS/TURQUOISE"	106
S73113	Symphysodon	aequifasciatus	DISCUS "TÜRKIS/TURQUOISE"	107
S73114	Symphysodon	aequifasciatus	DISCUS "TÜRKIS/TURQUOISE"	107
S73115	Symphysodon	aequifasciatus	DISCUS "TÜRKIS/TURQUOISE"	107
S73116	Symphysodon	aequifasciatus	DISCUS "TÜRKIS/TURQUOISE"	107
S73117	Symphysodon	aequifasciatus	DISCUS "TÜRKIS/TURQUOISE"	107
S73118	Symphysodon	aequifasciatus	DISCUS "TÜRKIS/TURQUOISE"	107
S73119	Symphysodon	aequifasciatus	DISCUS "TÜRKIS/TURQUOISE"	108/10
S73120	Symphysodon	aequifasciatus	DISCUS "TÜRKIS/TURQUOISE"	108
S73121	Symphysodon	aequifasciatus	DISCUS "TÜRKIS/TURQUOISE"	108
S73122	Symphysodon	aequifasciatus	DISCUS "TÜRKIS/TURQUOISE"	108
S73123	Symphysodon	aequifasciatus	DISCUS "TÜRKIS/TURQUOISE"	108
S73124	Symphysodon	aequifasciatus	DISCUS "TÜRKIS/TURQUOISE"	108
S73125	Symphysodon	aequifasciatus	DISCUS "TÜRKIS/TURQUOISE"	109
S73126	Symphysodon	aequifasciatus	DISCUS "TÜRKIS/TURQUOISE"	109
S73127	Symphysodon	aequifasciatus	DISCUS "TÜRKIS/TURQUOISE"	109
S73128	Symphysodon	aequifasciatus	DISCUS "TÜRKIS/TURQUOISE"	109
S73129	Symphysodon	aequifasciatus	DISCUS "TÜRKIS/TURQUOISE"	109
S73130	Symphysodon	aequifasciatus	DISCUS "TÜRKIS/TURQUOISE"	109
S73131	Symphysodon	aequifasciatus	DISCUS "TÜRKIS/TURQUOISE"	111
S73132	Symphysodon	aequifasciatus	DISCUS "TÜRKIS/TURQUOISE"	111
S73133	Symphysodon	aequifasciatus	DISCUS "TÜRKIS/TURQUOISE"	111
S73134	Symphysodon	aequifasciatus	DISCUS "TÜRKIS/TURQUOISE"	111
S73135	Symphysodon	aequifasciatus	DISCUS "TÜRKIS/TURQUOISE"	111
S73136	Symphysodon	aequifasciatus	DISCUS "TÜRKIS/TURQUOISE"	111
S73137	Symphysodon	aequifasciatus	DISCUS "TÜRKIS/TURQUOISE"	112
S73138	Symphysodon	aequifasciatus	DISCUS "TÜRKIS/TURQUOISE"	112
S73139	Symphysodon	aequifasciatus	DISCUS "TÜRKIS/TURQUOISE"	112
S73140	Symphysodon	aequifasciatus	DISCUS "TÜRKIS/TURQUOISE"	112
S73141	Symphysodon	aequifasciatus	DISCUS "TÜRKIS/TURQUOISE"	112
S73142	Symphysodon	aequifasciatus	DISCUS "TÜRKIS/TURQUOISE"	112
S73143	Symphysodon	aequifasciatus	DISCUS "TÜRKIS/TURQUOISE"	113
S73144	Symphysodon	aequifasciatus	DISCUS "TÜRKIS/TURQUOISE"	113/15
S73145	Symphysodon	aequifasciatus	DISCUS "TÜRKIS/TURQUOISE"	113
S73146	Symphysodon	aequifasciatus	DISCUS "TÜRKIS/TURQUOISE"	113

INDEX
Code - Nummern / Code - numbers

Code	Genus	Species	Common name	Page
S73147	Symphysodon	aequifasciatus	DISCUS "TÜRKIS/TURQUOISE"	113
S73148	Symphysodon	aequifasciatus	DISCUS "TÜRKIS/TURQUOISE"	113
S73149	Symphysodon	aequifasciatus	DISCUS "TÜRKIS/TURQUOISE"	114
S73150	Symphysodon	aequifasciatus	DISCUS "TÜRKIS/TURQUOISE"	114
S73151	Symphysodon	aequifasciatus	DISCUS "TÜRKIS/TURQUOISE"	114
S73152	Symphysodon	aequifasciatus	DISCUS "TÜRKIS/TURQUOISE"	114
S73153	Symphysodon	aequifasciatus	DISCUS "TÜRKIS/TURQUOISE"	114
S73154	Symphysodon	aequifasciatus	DISCUS "TÜRKIS/TURQUOISE"	114/16
S73155	Symphysodon	aequifasciatus	DISCUS "TÜRKIS/TURQUOISE"	117
S73156	Symphysodon	aequifasciatus	DISCUS "TÜRKIS/TURQUOISE"	117
S73157	Symphysodon	aequifasciatus	DISCUS "TÜRKIS/TURQUOISE"	117
S73158	Symphysodon	aequifasciatus	DISCUS "TÜRKIS/TURQUOISE"	117
S73159	Symphysodon	aequifasciatus	DISCUS "TÜRKIS/TURQUOISE"	117
S73160	Symphysodon	aequifasciatus	DISCUS "TÜRKIS/TURQUOISE"	117
S73161	Symphysodon	aequifasciatus	DISCUS "TÜRKIS/TURQUOISE"	118
S73162	Symphysodon	aequifasciatus	DISCUS "TÜRKIS/TURQUOISE"	118
S73163	Symphysodon	aequifasciatus	DISCUS "TÜRKIS/TURQUOISE"	118
S73164	Symphysodon	aequifasciatus	DISCUS "TÜRKIS/TURQUOISE"	118
S73165	Symphysodon	aequifasciatus	DISCUS "TÜRKIS/TURQUOISE"	118
S73166	Symphysodon	aequifasciatus	DISCUS "TÜRKIS/TURQUOISE"	118
S73167	Symphysodon	aequifasciatus	DISCUS "TÜRKIS/TURQUOISE"	120
S73168	Symphysodon	aequifasciatus	DISCUS "TÜRKIS/TURQUOISE"	120
S73169	Symphysodon	aequifasciatus	DISCUS "TÜRKIS/TURQUOISE"	120
S73170	Symphysodon	aequifasciatus	DISCUS "TÜRKIS/TURQUOISE"	120
S73171	Symphysodon	aequifasciatus	DISCUS "TÜRKIS/TURQUOISE"	120
S73172	Symphysodon	aequifasciatus	DISCUS "TÜRKIS/TURQUOISE"	120
S73173	Symphysodon	aequifasciatus	DISCUS "TÜRKIS/TURQUOISE"	119
S73179	Symphysodon	aequifasciatus	DISCUS "TÜRKIS/TURQUOISE"	121
S73180	Symphysodon	aequifasciatus	DISCUS "TÜRKIS/TURQUOISE"	121
S73181	Symphysodon	aequifasciatus	DISCUS "TÜRKIS/TURQUOISE"	121
S73182	Symphysodon	aequifasciatus	DISCUS "TÜRKIS/TURQUOISE"	121
S73183	Symphysodon	aequifasciatus	DISCUS "TÜRKIS/TURQUOISE"	121
S73184	Symphysodon	aequifasciatus	DISCUS "TÜRKIS/TURQUOISE"	121
S73185	Symphysodon	aequifasciatus	DISCUS "TÜRKIS/TURQUOISE"	122
S73186	Symphysodon	aequifasciatus	DISCUS "TÜRKIS/TURQUOISE"	122
S73187	Symphysodon	aequifasciatus	DISCUS "TÜRKIS/TURQUOISE"	122
S73188	Symphysodon	aequifasciatus	DISCUS "TÜRKIS/TURQUOISE"	122
S73189	Symphysodon	aequifasciatus	DISCUS "TÜRKIS/TURQUOISE"	122
S73190	Symphysodon	aequifasciatus	DISCUS "TÜRKIS/TURQUOISE"	122
S73192	Symphysodon	aequifasciatus	DISCUS "TÜRKIS/TURQUOISE"	123
S73194	Symphysodon	aequifasciatus	DISCUS "TÜRKIS/TURQUOISE"	123
S73195	Symphysodon	aequifasciatus	DISCUS "TÜRKIS/TURQUOISE"	123
S73197	Symphysodon	aequifasciatus	DISCUS "TÜRKIS/TURQUOISE"	124
S73198	Symphysodon	aequifasciatus	DISCUS "TÜRKIS/TURQUOISE"	124
S73199	Symphysodon	aequifasciatus	DISCUS "TÜRKIS/TURQUOISE"	124
S73200	Symphysodon	aequifasciatus	DISCUS "TÜRKIS/TURQUOISE"	124
S73201	Symphysodon	aequifasciatus	DISCUS "TÜRKIS/TURQUOISE"	124
S73202	Symphysodon	aequifasciatus	DISCUS "TÜRKIS/TURQUOISE"	124
S73203	Symphysodon	aequifasciatus	DISCUS "TÜRKIS/TURQUOISE"	125
S73204	Symphysodon	aequifasciatus	DISCUS "TÜRKIS/TURQUOISE"	125
S73205	Symphysodon	aequifasciatus	DISCUS "TÜRKIS/TURQUOISE"	125
S73206	Symphysodon	aequifasciatus	DISCUS "TÜRKIS/TURQUOISE"	125
S73207	Symphysodon	aequifasciatus	DISCUS "TÜRKIS/TURQUOISE" (mit Gelege/with eggs)	125
S73208	Symphysodon	aequifasciatus	DISCUS "TÜRKIS/TURQUOISE"	125
S73209	Symphysodon	aequifasciatus	DISCUS "TÜRKIS/TURQUOISE"	126
S73210	Symphysodon	aequifasciatus	DISCUS "TÜRKIS/TURQUOISE"	126
S73211	Symphysodon	aequifasciatus	DISCUS "TÜRKIS/TURQUOISE"	126
S73212	Symphysodon	aequifasciatus	DISCUS "TÜRKIS/TURQUOISE"	126
S73213	Symphysodon	aequifasciatus	DISCUS "TÜRKIS/TURQUOISE"	126
S73214	Symphysodon	aequifasciatus	DISCUS "TÜRKIS/TURQUOISE"	126
S74001	Symphysodon	aequifasciatus	DISCUS "ROT-TÜRKIS / RED-TURQUOISE"	137
S74002	Symphysodon	aequifasciatus	DISCUS "ROT-TÜRKIS / RED-TURQUOISE"	137
S74003	Symphysodon	aequifasciatus	DISCUS "ROT-TÜRKIS / RED-TURQUOISE"	137
S74004	Symphysodon	aequifasciatus	DISCUS "ROT-TÜRKIS / RED-TURQUOISE"	137
S74005	Symphysodon	aequifasciatus	DISCUS "ROT-TÜRKIS / RED-TURQUOISE"	137
S74006	Symphysodon	aequifasciatus	DISCUS "ROT-TÜRKIS / RED-TURQUOISE"	137
S74007	Symphysodon	aequifasciatus	DISCUS "ROT-TÜRKIS / RED-TURQUOISE"	138
S74008	Symphysodon	aequifasciatus	DISCUS "ROT-TÜRKIS / RED-TURQUOISE"	138
S74009	Symphysodon	aequifasciatus	DISCUS "ROT-TÜRKIS / RED-TURQUOISE"	138
S74010	Symphysodon	aequifasciatus	DISCUS "ROT-TÜRKIS / RED-TURQUOISE"	138
S74011	Symphysodon	aequifasciatus	DISCUS "ROT-TÜRKIS / RED-TURQUOISE"	138
S74012	Symphysodon	aequifasciatus	DISCUS "ROT-TÜRKIS / RED-TURQUOISE"	138
S74013	Symphysodon	aequifasciatus	DISCUS "ROT-TÜRKIS / RED-TURQUOISE"	139
S74014	Symphysodon	aequifasciatus	DISCUS "ROT-TÜRKIS / RED-TURQUOISE"	139
S74015	Symphysodon	aequifasciatus	DISCUS "ROT-TÜRKIS / RED-TURQUOISE"	139
S74016	Symphysodon	aequifasciatus	DISCUS "ROT-TÜRKIS / RED-TURQUOISE"	139
S74017	Symphysodon	aequifasciatus	DISCUS "ROT-TÜRKIS / RED-TURQUOISE"	139
S74018	Symphysodon	aequifasciatus	DISCUS "ROT-TÜRKIS / RED-TURQUOISE"	139
S74019	Symphysodon	aequifasciatus	DISCUS "ROT-TÜRKIS / RED-TURQUOISE"	140
S74020	Symphysodon	aequifasciatus	DISCUS "ROT-TÜRKIS / RED-TURQUOISE"	140
S74021	Symphysodon	aequifasciatus	DISCUS "ROT-TÜRKIS / RED-TURQUOISE"	140
S74022	Symphysodon	aequifasciatus	DISCUS "ROT-TÜRKIS / RED-TURQUOISE"	140
S74023	Symphysodon	aequifasciatus	DISCUS "ROT-TÜRKIS / RED-TURQUOISE"	140
S74024	Symphysodon	aequifasciatus	DISCUS "ROT-TÜRKIS / RED-TURQUOISE"	140
S74025	Symphysodon	aequifasciatus	DISCUS "ROT-TÜRKIS / RED-TURQUOISE"	142
S74027	Symphysodon	aequifasciatus	DISCUS "ROT-TÜRKIS / RED-TURQUOISE"	142
S74028	Symphysodon	aequifasciatus	DISCUS "ROT-TÜRKIS / RED-TURQUOISE"	142
S74029	Symphysodon	aequifasciatus	DISCUS "ROT-TÜRKIS / RED-TURQUOISE" (aus Royal-Blue)	142
S74030	Symphysodon	aequifasciatus	DISCUS "ROT-TÜRKIS / RED-TURQUOISE" (aus Royal-Blue)	142/44
S74031	Symphysodon	aequifasciatus	DISCUS "ROT-TÜRKIS / RED-TURQUOISE"	143
S74032	Symphysodon	aequifasciatus	DISCUS "ROT-TÜRKIS / RED-TURQUOISE"	143
S74033	Symphysodon	aequifasciatus	DISCUS "ROT-TÜRKIS / RED-TURQUOISE"	143/45
S74034	Symphysodon	aequifasciatus	DISCUS "ROT-TÜRKIS / RED-TURQUOISE"	143
S74035	Symphysodon	aequifasciatus	DISCUS "ROT-TÜRKIS / RED-TURQUOISE"	143
S74036	Symphysodon	aequifasciatus	DISCUS "ROT-TÜRKIS / RED-TURQUOISE"	143
S74037	Symphysodon	aequifasciatus	DISCUS "ROT-TÜRKIS / RED-TURQUOISE"	146
S74038	Symphysodon	aequifasciatus	DISCUS "ROT-TÜRKIS / RED-TURQUOISE"	146
S74039	Symphysodon	aequifasciatus	DISCUS "ROT-TÜRKIS / RED-TURQUOISE"	146
S74040	Symphysodon	aequifasciatus	DISCUS "ROT-TÜRKIS / RED-TURQUOISE"	146
S74041	Symphysodon	aequifasciatus	DISCUS "ROT-TÜRKIS / RED-TURQUOISE"	146
S74042	Symphysodon	aequifasciatus	DISCUS "ROT-TÜRKIS / RED-TURQUOISE"	146
S74043	Symphysodon	aequifasciatus	DISCUS "ROT-TÜRKIS / RED-TURQUOISE"	147
S74044	Symphysodon	aequifasciatus	DISCUS "ROT-TÜRKIS / RED-TURQUOISE"	147
S74045	Symphysodon	aequifasciatus	DISCUS "ROT-TÜRKIS / RED-TURQUOISE"	147
S74046	Symphysodon	aequifasciatus	DISCUS "ROT-TÜRKIS / RED-TURQUOISE"	147
S74047	Symphysodon	aequifasciatus	DISCUS "ROT-TÜRKIS / RED-TURQUOISE" Pair (aus Royal-Blue F1)	147
S74048	Symphysodon	aequifasciatus	DISCUS "ROT-TÜRKIS / RED-TURQUOISE" Pair	147
S74049	Symphysodon	aequifasciatus	DISCUS "ROT-TÜRKIS / RED-TURQUOISE"	148
S74050	Symphysodon	aequifasciatus	DISCUS "ROT-TÜRKIS / RED-TURQUOISE"	148
S74051	Symphysodon	aequifasciatus	DISCUS "ROT-TÜRKIS / RED-TURQUOISE"	148
S74052	Symphysodon	aequifasciatus	DISCUS "ROT-TÜRKIS / RED-TURQUOISE"	148
S74053	Symphysodon	aequifasciatus	DISCUS "ROT-TÜRKIS / RED-TURQUOISE"	148
S74054	Symphysodon	aequifasciatus	DISCUS "ROT-TÜRKIS / RED-TURQUOISE"	148
S74055	Symphysodon	aequifasciatus	DISCUS "ROT-TÜRKIS / RED-TURQUOISE"	149
S74056	Symphysodon	aequifasciatus	DISCUS "ROT-TÜRKIS / RED-TURQUOISE"	149
S74057	Symphysodon	aequifasciatus	DISCUS "ROT-TÜRKIS / RED-TURQUOISE"	149
S74058	Symphysodon	aequifasciatus	DISCUS "ROT-TÜRKIS / RED-TURQUOISE"	149
S74059	Symphysodon	aequifasciatus	DISCUS "ROT-TÜRKIS / RED-TURQUOISE"	149
S74060	Symphysodon	aequifasciatus	DISCUS "ROT-TÜRKIS / RED-TURQUOISE"	149
S74061	Symphysodon	aequifasciatus	DISCUS "ROT-TÜRKIS / RED-TURQUOISE"	150
S74062	Symphysodon	aequifasciatus	DISCUS "ROT-TÜRKIS / RED-TURQUOISE"	150
S74063	Symphysodon	aequifasciatus	DISCUS "ROT-TÜRKIS / RED-TURQUOISE"	150
S74064	Symphysodon	aequifasciatus	DISCUS "ROT-TÜRKIS / RED-TURQUOISE"	150
S74065	Symphysodon	aequifasciatus	DISCUS "ROT-TÜRKIS / RED-TURQUOISE"	150
S74066	Symphysodon	aequifasciatus	DISCUS "ROT-TÜRKIS / RED-TURQUOISE"	150
S74067	Symphysodon	aequifasciatus	DISCUS "ROT-TÜRKIS / RED-TURQUOISE"	151
S74068	Symphysodon	aequifasciatus	DISCUS "ROT-TÜRKIS / RED-TURQUOISE"	151

INDEX
Code - Nummern / Code - numbers

Code	Genus	Species	Common Name	Page
S74069	Symphysodon	aequifasciatus	DISCUS "ROT-TÜRKIS / RED-TURQUOISE"	151
S74070	Symphysodon	aequifasciatus	DISCUS "ROT-TÜRKIS / RED-TURQUOISE"	151
S74071	Symphysodon	aequifasciatus	DISCUS "ROT-TÜRKIS / RED-TURQUOISE"	151
S74072	Symphysodon	aequifasciatus	DISCUS "ROT-TÜRKIS / RED-TURQUOISE"	151
S74073	Symphysodon	aequifasciatus	DISCUS "ROT-TÜRKIS / RED-TURQUOISE"	152
S74074	Symphysodon	aequifasciatus	DISCUS "ROT-TÜRKIS / RED-TURQUOISE"	152
S74075	Symphysodon	aequifasciatus	DISCUS "ROT-TÜRKIS / RED-TURQUOISE"	152
S74076	Symphysodon	aequifasciatus	DISCUS "ROT-TÜRKIS / RED-TURQUOISE"	152
S74077	Symphysodon	aequifasciatus	DISCUS "ROT-TÜRKIS / RED-TURQUOISE"	152
S74078	Symphysodon	aequifasciatus	DISCUS "ROT-TÜRKIS / RED-TURQUOISE"	152
S74079	Symphysodon	aequifasciatus	DISCUS "ROT-TÜRKIS / RED-TURQUOISE"	153
S74082	Symphysodon	aequifasciatus	DISCUS "ROT-TÜRKIS / RED-TURQUOISE"	141
S74083	Symphysodon	aequifasciatus	DISCUS "ROT-TÜRKIS / RED-TURQUOISE"	141
S74084	Symphysodon	aequifasciatus	DISCUS "ROT-TÜRKIS / RED-TURQUOISE"	141
S74085	Symphysodon	aequifasciatus	DISCUS "ROT-TÜRKIS / RED-TURQUOISE"	141
S74086	Symphysodon	aequifasciatus	DISCUS "ROT-TÜRKIS / RED-TURQUOISE"	141
S74087	Symphysodon	aequifasciatus	DISCUS "ROT-TÜRKIS / RED-TURQUOISE"	141
S74088	Symphysodon	aequifasciatus	DISCUS "ROT-TÜRKIS / RED-TURQUOISE"	155
S74089	Symphysodon	aequifasciatus	DISCUS "ROT-TÜRKIS / RED-TURQUOISE"	155
S74090	Symphysodon	aequifasciatus	DISCUS "ROT-TÜRKIS / RED-TURQUOISE"	155
S74091	Symphysodon	aequifasciatus	DISCUS "ROT-TÜRKIS / RED-TURQUOISE"	155
S74092	Symphysodon	aequifasciatus	DISCUS "ROT-TÜRKIS / RED-TURQUOISE"	155
S74093	Symphysodon	aequifasciatus	DISCUS "ROT-TÜRKIS / RED-TURQUOISE"	155
S74094	Symphysodon	aequifasciatus	DISCUS "ROT-TÜRKIS / RED-TURQUOISE"	158
S74095	Symphysodon	aequifasciatus	DISCUS "ROT-TÜRKIS / RED-TURQUOISE"	158
S74096	Symphysodon	aequifasciatus	DISCUS "ROT-TÜRKIS / RED-TURQUOISE"	158/60
S74097	Symphysodon	aequifasciatus	DISCUS "ROT-TÜRKIS / RED-TURQUOISE"	158
S74098	Symphysodon	aequifasciatus	DISCUS "ROT-TÜRKIS / RED-TURQUOISE"	158
S74099	Symphysodon	aequifasciatus	DISCUS "ROT-TÜRKIS / RED-TURQUOISE"	158
S74100	Symphysodon	aequifasciatus	DISCUS "ROT-TÜRKIS / RED-TURQUOISE" "Tangerine Dream F1"	159
S74101	Symphysodon	aequifasciatus	DISCUS "ROT-TÜRKIS / RED-TURQUOISE" "Tangerine Dream F2"	159
S74102	Symphysodon	aequifasciatus	DISCUS "ROT-TÜRKIS / RED-TURQUOISE" "Tangerine Dream F3"	159
S74103	Symphysodon	aequifasciatus	DISCUS "ROT-TÜRKIS / RED-TURQUOISE" "Red Silk F1"	161
S74104	Symphysodon	aequifasciatus	DISCUS "ROT-TÜRKIS / RED-TURQUOISE" "Red Silk F2"	161
S74105	Symphysodon	aequifasciatus	DISCUS "ROT-TÜRKIS / RED-TURQUOISE" "Red Silk F3"	162
S74106	Symphysodon	aequifasciatus	DISCUS "ROT-TÜRKIS / RED-TURQUOISE" "Red Silk F3"	162
S74107	Symphysodon	aequifasciatus	DISCUS "ROT-TÜRKIS / RED-TURQUOISE" "Red Silk F3"	162
S74108	Symphysodon	aequifasciatus	DISCUS "ROT-TÜRKIS / RED-TURQUOISE" "Red Silk F3"	162
S74109	Symphysodon	aequifasciatus	DISCUS "ROT-TÜRKIS / RED-TURQUOISE" "Red Silk F3"	162
S74110	Symphysodon	aequifasciatus	DISCUS "ROT-TÜRKIS / RED-TURQUOISE" "Red Silk F3"	162/64
S74111	Symphysodon	aequifasciatus	DISCUS "ROT-TÜRKIS / RED-TURQUOISE"	154
S74112	Symphysodon	aequifasciatus	DISCUS "ROT-TÜRKIS / RED-TURQUOISE"	154
S74113	Symphysodon	aequifasciatus	DISCUS "ROT-TÜRKIS / RED-TURQUOISE"	154
S74114	Symphysodon	aequifasciatus	DISCUS "ROT-TÜRKIS / RED-TURQUOISE"	154
S74115	Symphysodon	aequifasciatus	DISCUS "ROT-TÜRKIS / RED-TURQUOISE"	154
S74116	Symphysodon	aequifasciatus	DISCUS "ROT-TÜRKIS / RED-TURQUOISE"	154
S74120	Symphysodon	aequifasciatus	DISCUS "ROT-TÜRKIS / RED-TURQUOISE"	153
S74121	Symphysodon	aequifasciatus	DISCUS "ROT-TÜRKIS / RED-TURQUOISE"	153
S74122	Symphysodon	aequifasciatus	DISCUS "ROT-TÜRKIS / RED-TURQUOISE" (Sieger/Best of show Aquarama 1993)	153
S74123	Symphysodon	aequifasciatus	DISCUS "ROT-TÜRKIS / RED-TURQUOISE"	153
S74124	Symphysodon	aequifasciatus	DISCUS "ROT-TÜRKIS / RED-TURQUOISE"	153
S74130	Symphysodon	aequifasciatus	DISCUS "ROT-TÜRKIS / RED-TURQUOISE" "Tangerine Dream F1"	159
S74131	Symphysodon	aequifasciatus	DISCUS "ROT-TÜRKIS / RED-TURQUOISE" "Tangerine Dream F3"	159
S74132	Symphysodon	aequifasciatus	DISCUS "ROT-TÜRKIS / RED-TURQUOISE" "Tangerine Dream F3"	159/166
S74133	Symphysodon	aequifasciatus	DISCUS "ROT-TÜRKIS / RED-TURQUOISE" "Red Silk F1"	161
S74134	Symphysodon	aequifasciatus	DISCUS "ROT-TÜRKIS / RED-TURQUOISE" "Red Silk F1"	161
S74135	Symphysodon	aequifasciatus	DISCUS "ROT-TÜRKIS / RED-TURQUOISE" "Red Silk F1"	161
S74136	Symphysodon	aequifasciatus	DISCUS "ROT-TÜRKIS / RED-TURQUOISE" "Red Silk F2"	161
S74150	Symphysodon	aequifasciatus	DISCUS "ROT-TÜRKIS / RED-TURQUOISE"	156
S74151	Symphysodon	aequifasciatus	DISCUS "ROT-TÜRKIS / RED-TURQUOISE"	156
S74152	Symphysodon	aequifasciatus	DISCUS "ROT-TÜRKIS / RED-TURQUOISE"	156
S74153	Symphysodon	aequifasciatus	DISCUS "ROT-TÜRKIS / RED-TURQUOISE"	156
S74154	Symphysodon	aequifasciatus	DISCUS "ROT-TÜRKIS / RED-TURQUOISE"	156
S74155	Symphysodon	aequifasciatus	DISCUS "ROT-TÜRKIS / RED-TURQUOISE"	156
S74156	Symphysodon	aequifasciatus	DISCUS "ROT-TÜRKIS / RED-TURQUOISE"	157
S75001	Symphysodon	aequifasciatus	DISCUS "rot geperlt / PEARL-RED"	168
S75002	Symphysodon	aequifasciatus	DISCUS "rot geperlt / PEARL-RED"	168
S75003	Symphysodon	aequifasciatus	DISCUS "GIANT PEARL"	169
S75004	Symphysodon	aequifasciatus	DISCUS "JEWEL OF MAYANS"	169
S75006	Symphysodon	aequifasciatus	DISCUS "BLUE JEWEL"	169
S75007	Symphysodon	aequifasciatus	DISCUS "BLUE PEARL JEWEL"	169
S75008	Symphysodon	aequifasciatus	DISCUS "RED SPOTTED JEWEL"	169
S75009	Symphysodon	aequifasciatus	DISCUS "CHOCOLATE PEARL" (open Champion Aquarama 1995)	170
S75010	Symphysodon	aequifasciatus	DISCUS "GODWIN SIM"	170
S75011	Symphysodon	aequifasciatus	DISCUS "BLOOD PEARL"	170
S75012	Symphysodon	aequifasciatus	DISCUS "BLOOD PEARL" -big spots-	170
S75013	Symphysodon	aequifasciatus	DISCUS "BLOOD PEARL" -red body-	170
S75015	Symphysodon	aequifasciatus	DISCUS "PEARL HIGHFIN"	171
S75016	Symphysodon	aequifasciatus	DISCUS "STRIPED HIGHFIN"	171
S75017	Symphysodon	aequifasciatus	DISCUS "offene Klasse / open class"	172
S75018	Symphysodon	aequifasciatus	DISCUS "offene Klasse / open class"	172
S75019	Symphysodon	aequifasciatus	DISCUS "offene Klasse / open class"	172
S75020	Symphysodon	aequifasciatus	DISCUS "offene Klasse / open class"	172
S75021	Symphysodon	aequifasciatus	DISCUS "offene Klasse / open class"	172
S75022	Symphysodon	aequifasciatus	DISCUS "offene Klasse / open class"	172
S75023	Symphysodon	aequifasciatus	DISCUS "offene Klasse / open class" "GHOST"	173
S75024	Symphysodon	aequifasciatus	DISCUS "offene Klasse / open class"	173
S75025	Symphysodon	aequifasciatus	DISCUS "offene Klasse / open class" "GHOST"	173
S75026	Symphysodon	aequifasciatus	DISCUS "offene Klasse / open class"	174
S75027	Symphysodon	aequifasciatus	DISCUS "offene Klasse / open class" "PIDGEON BLOOD"	175
S75029	Symphysodon	aequifasciatus	DISCUS "offene Klasse / open class" "PIDGEON BLOOD"	175
S75030	Symphysodon	aequifasciatus	DISCUS "offene Klasse / open class" "PIDGEON BLOOD"	175
S75031	Symphysodon	aequifasciatus	DISCUS "offene Klasse / open class" "PIDGEON BLOOD"	175
S75032	Symphysodon	aequifasciatus	DISCUS "offene Klasse / open class" "PIDGEON BLOOD"	175
S75033	Symphysodon	aequifasciatus	DISCUS "offene Klasse / open class" "PIDGEON BLOOD"	176
S75034	Symphysodon	aequifasciatus	DISCUS "offene Klasse / open class" "PIDGEON BLOOD"	176
S75035	Symphysodon	aequifasciatus	DISCUS "offene Klasse / open class" "PIDGEON BLOOD"	176
S75036	Symphysodon	aequifasciatus	DISCUS "offene Klasse / open class" "PIDGEON BLOOD"	176
S75037	Symphysodon	aequifasciatus	DISCUS "offene Klasse / open class" "PIDGEON BLOOD"	176
S75038	Symphysodon	aequifasciatus	DISCUS "offene Klasse / open class" "PIDGEON BLOOD"	176
S75039	Symphysodon	aequifasciatus	DISCUS "offene Klasse / open class" "BLUE SAPHIRE"	177
S75040	Symphysodon	aequifasciatus	DISCUS "offene Klasse / open class" "PIDGEON BLOOD"	177
S75041	Symphysodon	aequifasciatus	DISCUS "offene Klasse / open class" "PIDGEON BLOOD"	177
S75042	Symphysodon	aequifasciatus	DISCUS "offene Klasse / open class" "PIDGEON BLOOD"	177
S75043	Symphysodon	aequifasciatus	DISCUS "offene Klasse / open class" "PIDGEON BLOOD"	177
S75044	Symphysodon	aequifasciatus	DISCUS "offene Klasse / open class" "PIDGEON BLOOD"	177
S75045	Symphysodon	aequifasciatus	DISCUS "CHECKERBOARD"	168
S75046	Symphysodon	aequifasciatus	DISCUS "SNAKE SKIN"	171
S75047	Symphysodon	aequifasciatus	DISCUS "GLASS FIN"	171
S75048	Symphysodon	aequifasciatus	DISCUS "SNAKE SKIN"	171
S75049	Symphysodon	aequifasciatus	DISCUS "SNAKE SKIN"	171
S75050	Symphysodon	aequifasciatus	DISCUS "offene Klasse / open class"	173
S75051	Symphysodon	aequifasciatus	DISCUS "offene Klasse / open class"	173
S75052	Symphysodon	aequifasciatus	DISCUS "offene Klasse / open class"	173
S75053	Symphysodon	aequifasciatus	DISCUS "offene Klasse / open class" "YELLOW RAINBOW"	174
S75054	Symphysodon	aequifasciatus	DISCUS "offene Klasse / open class" "SKYBLUE"	174
S75055	Symphysodon	aequifasciatus	DISCUS "offene Klasse / open class" "PIDGEON BLOOD"	178
S75060	Symphysodon	aequifasciatus	DISCUS "CHECKERBOARD"	168
S75061	Symphysodon	aequifasciatus	DISCUS "CHECKERBOARD"	168
S75062	Symphysodon	aequifasciatus	DISCUS "CHECKERBOARD"	168
S75063	Symphysodon	aequifasciatus	DISCUS "offene Klasse / open class" "GOLDEN GHOST"	174
S75064	Symphysodon	aequifasciatus	DISCUS "offene Klasse / open class" "GOLDEN GHOST"	174
S89300	Symphysodon	aequifasciatus	DISCUS BROWN "Rio Madeira" (Gewinner/Best of show 1996/Duisburg)	47
S89301	Symphysodon	aequifasciatus	DISCUS BROWN "Rio Madeira"	42
S89302	Symphysodon	aequifasciatus	DISCUS BROWN "Rio Madeira"	42
S89303	Symphysodon	aequifasciatus	DISCUS BROWN "Rio Madeira"	42
S89304	Symphysodon	aequifasciatus	DISCUS BROWN "Rio Madeira"	42

INDEX
Code - Nummern / Code - numbers

Code	Genus	Species	Name	Page
S89305	Symphysodon	aequifasciatus	DISCUS BROWN "Rio Madeira"	42
S89306	Symphysodon	aequifasciatus	DISCUS BROWN "Rio Madeira"	42
S89307	Symphysodon	aequifasciatus	DISCUS BROWN "Rio Madeira"	43
S89308	Symphysodon	aequifasciatus	DISCUS BROWN "Rio Madeira"	43
S89309	Symphysodon	aequifasciatus	DISCUS BROWN "Rio Madeira"	43
S89310	Symphysodon	aequifasciatus	DISCUS BROWN "Rio Madeira"	43
S89311	Symphysodon	aequifasciatus	DISCUS BROWN "Rio Madeira"	43
S89312	Symphysodon	aequifasciatus	DISCUS BROWN "Rio Madeira"	43
S89313	Symphysodon	aequifasciatus	DISCUS BROWN "Rio Madeira"	44
S89314	Symphysodon	aequifasciatus	DISCUS BROWN "Rio Madeira"	44
S89315	Symphysodon	aequifasciatus	DISCUS BROWN "Rio Madeira"	44
S89316	Symphysodon	aequifasciatus	DISCUS BROWN "Rio Madeira"	44
S89317	Symphysodon	aequifasciatus	DISCUS BROWN "Rio Madeira"	44
S89318	Symphysodon	aequifasciatus	DISCUS BROWN "Rio Madeira"	44
S89319	Symphysodon	aequifasciatus	DISCUS BROWN "Rio Madeira"	45
S89320	Symphysodon	aequifasciatus	DISCUS BROWN "Rio Madeira"	45
S89321	Symphysodon	aequifasciatus	DISCUS BROWN "Rio Madeira"	45
S89322	Symphysodon	aequifasciatus	DISCUS BROWN "Rio Madeira"	45
S89323	Symphysodon	aequifasciatus	DISCUS BROWN "Rio Madeira"	45
S89324	Symphysodon	aequifasciatus	DISCUS BROWN "Rio Madeira"	45
S89325	Symphysodon	aequifasciatus	DISCUS BROWN "Rio Madeira"	46
S89326	Symphysodon	aequifasciatus	DISCUS BROWN "Rio Madeira"	46
S89327	Symphysodon	aequifasciatus	DISCUS BROWN "Rio Madeira"	46
S89328	Symphysodon	aequifasciatus	DISCUS BROWN "Rio Madeira"	46
S89329	Symphysodon	aequifasciatus	DISCUS BROWN "Rio Madeira"	46
S89331	Symphysodon	aequifasciatus	DISCUS BROWN "Rio Madeira"	48
S89332	Symphysodon	aequifasciatus	DISCUS BROWN "Rio Madeira"	48
S89333	Symphysodon	aequifasciatus	DISCUS BROWN "Rio Madeira"	48
S89334	Symphysodon	aequifasciatus	DISCUS BROWN "Rio Madeira"	48
S89335	Symphysodon	aequifasciatus	DISCUS BROWN "Rio Madeira"	48
S89336	Symphysodon	aequifasciatus	DISCUS BROWN "Rio Madeira"	48
S89337	Symphysodon	aequifasciatus	DISCUS BROWN "Rio Madeira"	49
S89338	Symphysodon	aequifasciatus	DISCUS BROWN "Rio Madeira"	49
S89339	Symphysodon	aequifasciatus	DISCUS BROWN "Rio Madeira"	49
S89340	Symphysodon	aequifasciatus	DISCUS BROWN "Rio Madeira"	49
S89341	Symphysodon	aequifasciatus	DISCUS BROWN "Rio Madeira"	49
S89342	Symphysodon	aequifasciatus	DISCUS BROWN "Rio Madeira"	49
S89343	Symphysodon	aequifasciatus	DISCUS BROWN "Rio Madeira"	50
S89344	Symphysodon	aequifasciatus	DISCUS BROWN "Rio Madeira"	50
S89345	Symphysodon	aequifasciatus	DISCUS BROWN "Rio Madeira"	50
S89346	Symphysodon	aequifasciatus	DISCUS BROWN "Rio Madeira"	50
S89347	Symphysodon	aequifasciatus	DISCUS BROWN "Rio Madeira"	50
S89348	Symphysodon	aequifasciatus	DISCUS BROWN "Rio Madeira"	50
S89349	Symphysodon	aequifasciatus	DISCUS BROWN "Rio Madeira"	51
S89350	Symphysodon	aequifasciatus	DISCUS BROWN "Rio Madeira"	51
S89351	Symphysodon	aequifasciatus	DISCUS BROWN "Rio Madeira"	51
S89352	Symphysodon	aequifasciatus	DISCUS BROWN "Rio Madeira"	51
S89353	Symphysodon	aequifasciatus	DISCUS BROWN "Rio Madeira"	51
S89354	Symphysodon	aequifasciatus	DISCUS BROWN "Rio Madeira"	52
S89355	Symphysodon	aequifasciatus	DISCUS BROWN "Rio Madeira"	52
S89356	Symphysodon	aequifasciatus	DISCUS BROWN "Rio Madeira"	52
S89357	Symphysodon	aequifasciatus	DISCUS BROWN "Rio Madeira"	52
S89358	Symphysodon	aequifasciatus	DISCUS BROWN "Rio Madeira"	52
S89359	Symphysodon	aequifasciatus	DISCUS BROWN "Rio Madeira"	52
S89360	Symphysodon	aequifasciatus	DISCUS BROWN "Rio Madeira"	52
S89361	Symphysodon	aequifasciatus	DISCUS BROWN "Rio Madeira"	53
S89362	Symphysodon	aequifasciatus	DISCUS BROWN "Rio Madeira"	53
S89363	Symphysodon	aequifasciatus	DISCUS BROWN "Rio Madeira"	53
S89364	Symphysodon	aequifasciatus	DISCUS BROWN "Rio Madeira"	53
S89365	Symphysodon	aequifasciatus	DISCUS BROWN "Rio Madeira"	53/55
S89366	Symphysodon	aequifasciatus	DISCUS BROWN "Rio Madeira"	53
S89367	Symphysodon	aequifasciatus	DISCUS BRAUN/BROWN	69
S89368	Symphysodon	aequifasciatus	DISCUS BRAUN/BROWN	69
S89369	Symphysodon	aequifasciatus	DISCUS BRAUN/BROWN	69
S89370	Symphysodon	aequifasciatus	DISCUS BRAUN/BROWN	69
S89371	Symphysodon	aequifasciatus	DISCUS BRAUN/BROWN	69
S89372	Symphysodon	aequifasciatus	DISCUS BRAUN/BROWN	69
S89373	Symphysodon	aequifasciatus	DISCUS BRAUN/BROWN	70
S89374	Symphysodon	aequifasciatus	DISCUS BRAUN/BROWN	70
S89375	Symphysodon	aequifasciatus	DISCUS BRAUN/BROWN	70
S89376	Symphysodon	aequifasciatus	DISCUS BRAUN/BROWN	70
S89377	Symphysodon	aequifasciatus	DISCUS BRAUN/BROWN	70
S89378	Symphysodon	aequifasciatus	DISCUS BRAUN/BROWN	70
S89379	Symphysodon	aequifasciatus	DISCUS BRAUN/BROWN	71
S89380	Symphysodon	aequifasciatus	DISCUS BRAUN/BROWN (Rio Purus)	71
S89381	Symphysodon	aequifasciatus	DISCUS BRAUN/BROWN (Rio Purus)	71
S89382	Symphysodon	aequifasciatus	DISCUS BRAUN/BROWN	71
S89383	Symphysodon	aequifasciatus	DISCUS BRAUN/BROWN (Rio Ipixuna)	71
S89384	Symphysodon	aequifasciatus	DISCUS BRAUN/BROWN (Rio Ipixuna)	71
S89385	Symphysodon	aequifasciatus	DISCUS BRAUN/BROWN	72
S89386	Symphysodon	aequifasciatus	DISCUS BRAUN/BROWN	72
S89387	Symphysodon	aequifasciatus	DISCUS BRAUN/BROWN	72
S89388	Symphysodon	aequifasciatus	DISCUS BRAUN/BROWN	72
S89389	Symphysodon	aequifasciatus	DISCUS BRAUN/BROWN	72
S89390	Symphysodon	aequifasciatus	DISCUS BRAUN/BROWN	72
S89401	Symphysodon	aequifasciatus	DISCUS BROWN "Alenquer"	54
S89402	Symphysodon	aequifasciatus	DISCUS BROWN "Alenquer"	54
S89403	Symphysodon	aequifasciatus	DISCUS BROWN "Alenquer"	54
S89404	Symphysodon	aequifasciatus	DISCUS BROWN "Alenquer" (PAIR)	54
S89405	Symphysodon	aequifasciatus	DISCUS BROWN "Alenquer"	54
S89406	Symphysodon	aequifasciatus	DISCUS BROWN "Alenquer"	54/56
S89407	Symphysodon	aequifasciatus	DISCUS BROWN "Alenquer"	57
S89408	Symphysodon	aequifasciatus	DISCUS BROWN "Alenquer"	57
S89409	Symphysodon	aequifasciatus	DISCUS BROWN "Alenquer"	57
S89410	Symphysodon	aequifasciatus	DISCUS BROWN "Alenquer"	57
S89411	Symphysodon	aequifasciatus	DISCUS BROWN "Alenquer"	57
S89412	Symphysodon	aequifasciatus	DISCUS flächig rot (gelb/braun) / solid red (yellow/brown)	57/59+166
S89413	Symphysodon	aequifasciatus	DISCUS BROWN "Alenquer"	58
S89414	Symphysodon	aequifasciatus	DISCUS BROWN "Alenquer"	58
S89415	Symphysodon	aequifasciatus	DISCUS BROWN "Alenquer"	58
S89416	Symphysodon	aequifasciatus	DISCUS BROWN "Alenquer"	58
S89417	Symphysodon	aequifasciatus	DISCUS BROWN "Alenquer"	58
S89418	Symphysodon	aequifasciatus	DISCUS BROWN "Alenquer"	58
S89419	Symphysodon	aequifasciatus	DISCUS BROWN "Alenquer" (PAIR)	60
S89420	Symphysodon	aequifasciatus	DISCUS BROWN "Alenquer" (PAIR)	60
S89421	Symphysodon	aequifasciatus	DISCUS BROWN "Alenquer"	60
S89422	Symphysodon	aequifasciatus	DISCUS BROWN "Alenquer"	60
S89423	Symphysodon	aequifasciatus	DISCUS BROWN "Alenquer"	60
S89424	Symphysodon	aequifasciatus	DISCUS BROWN "Alenquer"	60
S89425	Symphysodon	aequifasciatus	DISCUS BROWN "Alenquer"	61
S89426	Symphysodon	aequifasciatus	DISCUS BROWN "Alenquer"	61
S89427	Symphysodon	aequifasciatus	DISCUS BROWN "Alenquer"	61
S89428	Symphysodon	aequifasciatus	DISCUS BROWN "Alenquer"	61
S89429	Symphysodon	aequifasciatus	DISCUS BROWN "Alenquer"	61
S89430	Symphysodon	aequifasciatus	DISCUS BROWN "Alenquer" (mit Larven)	61
S89431	Symphysodon	aequifasciatus	DISCUS BROWN "Alenquer Red Eddy"	62
S89432	Symphysodon	aequifasciatus	DISCUS BROWN "Alenquer Red Eddy"	62
S89433	Symphysodon	aequifasciatus	DISCUS BROWN "Alenquer Red Eddy"	62
S89434	Symphysodon	aequifasciatus	DISCUS BROWN "Alenquer"	62
S89435	Symphysodon	aequifasciatus	DISCUS BROWN "Alenquer"	62
S89436	Symphysodon	aequifasciatus	DISCUS BROWN "Alenquer"	62
S89437	Symphysodon	aequifasciatus	DISCUS BROWN "Alenquer"	63
S89438	Symphysodon	aequifasciatus	DISCUS BROWN "Alenquer"	63
S89501	Symphysodon	aequifasciatus	DISCUS BLAU/BLUE	73
S89502	Symphysodon	aequifasciatus	DISCUS BLAU/BLUE	73
S89503	Symphysodon	aequifasciatus	DISCUS BLAU/BLUE	73
S89504	Symphysodon	aequifasciatus	DISCUS BLAU/BLUE	73

INDEX
Code - Nummern / Code - numbers

Code	Genus	Species	Common Name	Page
S89505	Symphysodon	aequifasciatus	DISCUS BLAU/BLUE	73
S89506	Symphysodon	aequifasciatus	DISCUS BLAU/BLUE	73
S89507	Symphysodon	aequifasciatus	DISCUS BLAU/BLUE	74
S89508	Symphysodon	aequifasciatus	DISCUS BLAU/BLUE	74
S89509	Symphysodon	aequifasciatus	DISCUS BLAU/BLUE	74
S89510	Symphysodon	aequifasciatus	DISCUS BLAU/BLUE	74
S89511	Symphysodon	aequifasciatus	DISCUS BLAU/BLUE	74
S89512	Symphysodon	aequifasciatus	DISCUS BLAU/BLUE	74
S89513	Symphysodon	aequifasciatus	DISCUS BLAU/BLUE	75
S89514	Symphysodon	aequifasciatus	DISCUS BLAU/BLUE	75
S89515	Symphysodon	aequifasciatus	DISCUS BLAU/BLUE (Rio Nhamunda?)	75
S89516	Symphysodon	aequifasciatus	DISCUS BLAU/BLUE (Rio Nhamunda?)	75
S89517	Symphysodon	aequifasciatus	DISCUS BLAU/BLUE (Rio Nhamunda?)	75
S89518	Symphysodon	aequifasciatus	DISCUS BLAU/BLUE (Rio Nhamunda?)	75
S89519	Symphysodon	aequifasciatus	DISCUS BLAU/BLUE	76
S89520	Symphysodon	aequifasciatus	DISCUS BLAU/BLUE	76
S89521	Symphysodon	aequifasciatus	DISCUS BLAU/BLUE	76
S89522	Symphysodon	aequifasciatus	DISCUS BLAU/BLUE	76
S89523	Symphysodon	aequifasciatus	DISCUS BLAU/BLUE	76/78
S89524	Symphysodon	aequifasciatus	DISCUS BLAU/BLUE "Manacapuru(?)"	76
S89601	Symphysodon	aequifasciatus	DISCUS "ROYAL-BLUE"	79
S89602	Symphysodon	aequifasciatus	DISCUS "ROYAL-BLUE"	79
S89603	Symphysodon	aequifasciatus	DISCUS "ROYAL-BLUE"	79
S89604	Symphysodon	aequifasciatus	DISCUS "ROYAL-BLUE"	79
S89605	Symphysodon	aequifasciatus	DISCUS "ROYAL-BLUE"	79
S89606	Symphysodon	aequifasciatus	DISCUS "ROYAL-BLUE"	79
S89610	Symphysodon	aequifasciatus	DISCUS "ROYAL-BLUE"	80
S89611	Symphysodon	aequifasciatus	DISCUS "ROYAL-BLUE"	80
S89612	Symphysodon	aequifasciatus	DISCUS "ROYAL-BLUE" (F1)	81
S89613	Symphysodon	aequifasciatus	DISCUS "ROYAL-BLUE" (F1)	81
S89614	Symphysodon	aequifasciatus	DISCUS "ROYAL-BLUE" (F1)	81
S89615	Symphysodon	aequifasciatus	DISCUS "ROYAL-BLUE" (F1)	81
S89616	Symphysodon	aequifasciatus	DISCUS "ROYAL-BLUE" (F1)	81
S89617	Symphysodon	aequifasciatus	DISCUS "ROYAL-BLUE" (F1)	81
S89618	Symphysodon	aequifasciatus	DISCUS "ROYAL-BLUE" (F1)	82
S89619	Symphysodon	aequifasciatus	DISCUS "ROYAL-BLUE" (F1)	82
S89620	Symphysodon	aequifasciatus	DISCUS "ROYAL-BLUE" (F1)	82
S89621	Symphysodon	aequifasciatus	DISCUS "ROYAL-BLUE" (F1)	82
S89622	Symphysodon	aequifasciatus	DISCUS "ROYAL-BLUE" (F1)	82
S89623	Symphysodon	aequifasciatus	DISCUS "ROYAL-BLUE" (F1)	82
S89624	Symphysodon	aequifasciatus	DISCUS "ROYAL-BLUE" (F2)	83
S89625	Symphysodon	aequifasciatus	DISCUS "ROYAL-BLUE" (F2)	83
S89626	Symphysodon	aequifasciatus	DISCUS "ROYAL-BLUE" (F2)	83
S89655	Symphysodon	aequifasciatus	DISCUS BLAU/BLUE "Manacapuru" (F1)	77
S89656	Symphysodon	aequifasciatus	DISCUS BLAU/BLUE "Manacapuru" (F1)	77
S89657	Symphysodon	aequifasciatus	DISCUS BLAU/BLUE "Manacapuru" (F2)	77
S89658	Symphysodon	aequifasciatus	DISCUS BLAU/BLUE "Manacapuru" (F2)	77
S89659	Symphysodon	aequifasciatus	DISCUS BLAU/BLUE "Manacapuru" (F2)	77/84
S89701	Symphysodon	discus	DISCUS HECKEL	33
S89702	Symphysodon	discus	DISCUS HECKEL	33
S89703	Symphysodon	discus	DISCUS HECKEL	33
S89704	Symphysodon	discus	DISCUS HECKEL	33
S89705	Symphysodon	discus	DISCUS HECKEL	33
S89706	Symphysodon	discus	DISCUS HECKEL	33
S89707	Symphysodon	discus	DISCUS HECKEL	34
S89708	Symphysodon	discus	DISCUS HECKEL	34
S89709	Symphysodon	discus	DISCUS HECKEL	34
S89710	Symphysodon	discus	DISCUS HECKEL	34
S89711	Symphysodon	discus	DISCUS HECKEL	34
S89712	Symphysodon	discus	DISCUS HECKEL	34
S89801	Symphysodon	discus	DISCUS HECKEL -Blaukopf/Bluehead	35
S89802	Symphysodon	discus	DISCUS HECKEL -Blaukopf/Bluehead	35
S89803	Symphysodon	discus	DISCUS HECKEL -Blaukopf/Bluehead	35
S89804	Symphysodon	discus	DISCUS HECKEL -Blaukopf/Bluehead	35
S89805	Symphysodon	discus	DISCUS HECKEL -Blaukopf/Bluehead	35
S89806	Symphysodon	discus	DISCUS HECKEL -Blaukopf/Bluehead	35
S89810	Symphysodon	discus	Heckel-Discus	36+38
S89811	Symphysodon	discus	Heckel-Discus	39
S89812	Symphysodon	discus	Heckel-Discus	39
S89813	Symphysodon	discus	Heckel-Discus	39
S89814	Symphysodon	discus	Heckel-Discus	39
S89815	Symphysodon	discus	Heckel-Discus	39
S89816	Symphysodon	discus	Heckel-Discus	39
S89817	Symphysodon	discus	Heckel-Discus	36/37
S89820	Symphysodon	Hybride	Discus HYBRIDE (M: Hybr.WF x RT) x (F: Hybr. aus Hybr.WF x RT)	40
S89825	Symphysodon	Hybride	Discus HYBRIDE (Male: Heckel) x (Female: red-turquoise)	41
S89901	Symphysodon	aequifasciatus	DISCUS GRÜN/GREEN	85
S89902	Symphysodon	aequifasciatus	DISCUS GRÜN/GREEN	85
S89903	Symphysodon	aequifasciatus	DISCUS GRÜN/GREEN	85
S89904	Symphysodon	aequifasciatus	DISCUS GRÜN/GREEN	85
S89905	Symphysodon	aequifasciatus	DISCUS GRÜN/GREEN	85
S89906	Symphysodon	aequifasciatus	DISCUS GRÜN/GREEN	85
S89907	Symphysodon	aequifasciatus	DISCUS GRÜN/GREEN "Rio Jurua(?)"	86
S89908	Symphysodon	aequifasciatus	DISCUS GRÜN/GREEN	86
S89909	Symphysodon	aequifasciatus	DISCUS GRÜN/GREEN	86
S89910	Symphysodon	aequifasciatus	DISCUS GRÜN/GREEN "Rio Jurua(?)"	86
S89911	Symphysodon	aequifasciatus	DISCUS GRÜN/GREEN	86
S89912	Symphysodon	aequifasciatus	DISCUS GRÜN/GREEN	86
S89913	Symphysodon	aequifasciatus	DISCUS GRÜN/GREEN	87
S89914	Symphysodon	aequifasciatus	DISCUS GRÜN/GREEN	87
S89915	Symphysodon	aequifasciatus	DISCUS GRÜN/GREEN	88
S89916	Symphysodon	aequifasciatus	DISCUS GRÜN/GREEN	88
S89917	Symphysodon	aequifasciatus	DISCUS GRÜN/GREEN	88
S89918	Symphysodon	aequifasciatus	DISCUS GRÜN/GREEN	88
S89919	Symphysodon	aequifasciatus	DISCUS GRÜN/GREEN	88
S89920	Symphysodon	aequifasciatus	DISCUS GRÜN/GREEN	88
S89921	Symphysodon	aequifasciatus	DISCUS GRÜN/GREEN	89
S89922	Symphysodon	aequifasciatus	DISCUS GRÜN/GREEN	89
S89923	Symphysodon	aequifasciatus	DISCUS GRÜN/GREEN	89
S89924	Symphysodon	aequifasciatus	DISCUS GRÜN/GREEN	89
S89925	Symphysodon	aequifasciatus	DISCUS GRÜN/GREEN	89
S89926	Symphysodon	aequifasciatus	DISCUS GRÜN/GREEN	89
S89927	Symphysodon	aequifasciatus	DISCUS GRÜN/GREEN	90
S89928	Symphysodon	aequifasciatus	DISCUS GRÜN/GREEN	90
S89929	Symphysodon	aequifasciatus	DISCUS GRÜN/GREEN	90
S89930	Symphysodon	aequifasciatus	DISCUS GRÜN/GREEN	90
S89931	Symphysodon	aequifasciatus	DISCUS GRÜN/GREEN	90
S89932	Symphysodon	aequifasciatus	DISCUS GRÜN/GREEN	90
S89933	Symphysodon	aequifasciatus	DISCUS GRÜN/GREEN	91
S89934	Symphysodon	aequifasciatus	DISCUS GRÜN/GREEN	91
S89935	Symphysodon	aequifasciatus	DISCUS GRÜN/GREEN	91
S89936	Symphysodon	aequifasciatus	DISCUS GRÜN/GREEN	91
S89937	Symphysodon	aequifasciatus	DISCUS GRÜN/GREEN	91
S89938	Symphysodon	aequifasciatus	DISCUS GRÜN/GREEN	91
S89939	Symphysodon	aequifasciatus	DISCUS GRÜN/GREEN	92
S89940	Symphysodon	aequifasciatus	DISCUS GRÜN/GREEN	92
S89941	Symphysodon	aequifasciatus	DISCUS GRÜN/GREEN	92
S89942	Symphysodon	aequifasciatus	DISCUS GRÜN/GREEN "Rio Putumajo(?)"	92
S89943	Symphysodon	aequifasciatus	DISCUS GRÜN/GREEN	92
S89944	Symphysodon	aequifasciatus	DISCUS GRÜN/GREEN	92/94
S89945	Symphysodon	aequifasciatus	DISCUS GRÜN/GREEN	93
S89947	Symphysodon	aequifasciatus	DISCUS GRÜN/GREEN	93
S89948	Symphysodon	aequifasciatus	DISCUS "ROYAL-GREEN"	93
S89951	Symphysodon	aequifasciatus	DISCUS GRÜN/GREEN	95
S89952	Symphysodon	aequifasciatus	DISCUS GRÜN/GREEN	95
S89953	Symphysodon	aequifasciatus	DISCUS GRÜN/GREEN "TEFE"	87
S89954	Symphysodon	aequifasciatus	DISCUS GRÜN/GREEN "TEFE"	87

INDEX
Code - Nummern / Code - numbers

Code	Genus	Species	Description	Page
S89955	Symphysodon	aequifasciatus	DISCUS GRÜN/GREEN	95
S89956	Symphysodon	aequifasciatus	DISCUS GRÜN/GREEN "TEFE"	87
S89957	Symphysodon	aequifasciatus	DISCUS GRÜN/GREEN	96
S89958	Symphysodon	aequifasciatus	DISCUS GRÜN/GREEN	96
S89959	Symphysodon	aequifasciatus	DISCUS GRÜN/GREEN "SUNSET-RED"	96
S89960	Symphysodon	aequifasciatus	DISCUS GRÜN/GREEN "NOBEL-RED"	96
S89961	Symphysodon	aequifasciatus	DISCUS GRÜN/GREEN	96
S89962	Symphysodon	aequifasciatus	DISCUS GRÜN/GREEN	96
S89963	Symphysodon	aequifasciatus	DISCUS GRÜN/GREEN "COARI"	97
S89964	Symphysodon	aequifasciatus	DISCUS GRÜN/GREEN "TEFE"	87
S89965	Symphysodon	aequifasciatus	DISCUS GRÜN/GREEN "COARI"	97
S89966	Symphysodon	aequifasciatus	DISCUS GRÜN/GREEN "COARI"	97
S89967	Symphysodon	aequifasciatus	DISCUS GRÜN/GREEN "COARI"	97
S89969	Symphysodon	aequifasciatus	DISCUS GRÜN/GREEN "COARI"	98
S90100	Symphysodon	aequifasciatus	DISCUS flächig rot (gelb/braun) / solid red (yellow/brown)	167
S90101	Symphysodon	aequifasciatus	DISCUS flächig rot (gelb/braun) / solid red (yellow/brown) "YELLOW"	163
S90102	Symphysodon	aequifasciatus	DISCUS flächig rot (gelb/braun) / solid red (yellow/brown) "GOLDEN"	163
S90103	Symphysodon	aequifasciatus	DISCUS flächig rot (gelb/braun) / solid red (yellow/brown) "GOLDEN"	163
S90104	Symphysodon	aequifasciatus	DISCUS flächig rot (gelb/braun) / solid red (yellow/brown) "GOLDEN-RAINBOW"	163
S90105	Symphysodon	aequifasciatus	DISCUS flächig rot (gelb/braun) / solid red (yellow/brown) "SWEETS TEMPTATION"	163
S90106	Symphysodon	aequifasciatus	DISCUS flächig rot (gelb/braun) / solid red (yellow/brown) "GOLDEN-PHOENIX"	163
S90107	Symphysodon	aequifasciatus	DISCUS flächig rot (gelb/braun) / solid red (yellow/brown) "RED ANGEL"	165
S90108	Symphysodon	aequifasciatus	DISCUS flächig rot (gelb/braun) / solid red (yellow/brown) "BLUE ANGEL"	165
S90109	Symphysodon	aequifasciatus	DISCUS flächig rot (gelb/braun) / solid red (yellow/brown) "RED STRIPEHEAD"	165
S90110	Symphysodon	aequifasciatus	DISCUS flächig rot (gelb/braun) / solid red (yellow/brown) "MARLBORO-RED"	165
S90111	Symphysodon	aequifasciatus	DISCUS flächig rot (gelb/braun) / solid red (yellow/brown) "TOMATO-RED"	165
S90112	Symphysodon	aequifasciatus	DISCUS flächig rot (gelb/braun) / solid red (yellow/brown) "MARLBORO-RED"	165
S90113	Symphysodon	aequifasciatus	DISCUS flächig rot (gelb/braun) / solid red (yellow/brown) "PIDGEON BLOOD"	166
S90114	Symphysodon	aequifasciatus	DISCUS flächig rot (gelb/braun) / solid red (yellow/brown) "RED SILK" F3	166
S90117	Symphysodon	aequifasciatus	DISCUS "ROT / RED" with stripes	166
S90118	Symphysodon	aequifasciatus	DISCUS flächig rot (gelb/braun) / solid red (yellow/brown) "Rio Madeira"	167
S90119	Symphysodon	aequifasciatus	DISCUS flächig rot (gelb/braun) / solid red (yellow/brown) "Rio Madeira" F1	167
S90201	Symphysodon	aequifasciatus	DISCUS BROWN "Alenquer Red Eddy" F1	63
S90202	Symphysodon	aequifasciatus	DISCUS BROWN "Alenquer Red Eddy" F1	63
S90203	Symphysodon	aequifasciatus	DISCUS BROWN "Alenquer Red Eddy" F2	63
S90204	Symphysodon	aequifasciatus	DISCUS BROWN "Alenquer Red Eddy" F2	63/65
S90205	Symphysodon	aequifasciatus	DISCUS BROWN "Alenquer Red Eddy" F2	64
S90206	Symphysodon	aequifasciatus	DISCUS BROWN "Alenquer Red Eddy" F2	64
S90207	Symphysodon	aequifasciatus	DISCUS BROWN "Alenquer Red Eddy" F3	64
S90208	Symphysodon	aequifasciatus	DISCUS BROWN "Alenquer Red Eddy" F3	64
S90209	Symphysodon	aequifasciatus	DISCUS BROWN "Alenquer Red Eddy" F3	64
S90210	Symphysodon	aequifasciatus	DISCUS BROWN "Alenquer Red Eddy" F3	64/66
S90211	Symphysodon	aequifasciatus	DISCUS BROWN "Alenquer Red Eddy" F4	67/68
S90212	Symphysodon	aequifasciatus	DISCUS BROWN "Alenquer Red Eddy" F4	67/68
S90213	Symphysodon	aequifasciatus	DISCUS BROWN "Alenquer Red Eddy" F4	67
S90214	Symphysodon	aequifasciatus	DISCUS flächig rot/red with stripes "ALENQUER, Red Eddy"	67
S90501	Symphysodon	aequifasciatus	DISCUS GREEN-BLUE flächig/solid	127
S90502	Symphysodon	aequifasciatus	DISCUS GREEN-BLUE flächig/solid	127
S90503	Symphysodon	aequifasciatus	DISCUS GREEN-BLUE flächig/solid	127
S90504	Symphysodon	aequifasciatus	DISCUS GREEN-BLUE flächig/solid	127
S90505	Symphysodon	aequifasciatus	DISCUS GREEN-BLUE flächig/solid	127
S90506	Symphysodon	aequifasciatus	DISCUS -SOLID TURQUOISE	128
S90507	Symphysodon	aequifasciatus	DISCUS -SOLID TURQUOISE	128
S90508	Symphysodon	aequifasciatus	DISCUS -SOLID TURQUOISE	128
S90509	Symphysodon	aequifasciatus	DISCUS "FLÄCHIG BLAU/TÜRKIS / SOLID BLUE/TURQUOISE"	129
S90510	Symphysodon	aequifasciatus	DISCUS "FLÄCHIG BLAU/TÜRKIS / SOLID BLUE/TURQUOISE"	129
S90511	Symphysodon	aequifasciatus	DISCUS "FLÄCHIG BLAU/TÜRKIS / SOLID BLUE/TURQUOISE"	129
S90512	Symphysodon	aequifasciatus	DISCUS "FLÄCHIG BLAU/TÜRKIS / SOLID BLUE/TURQUOISE"	129
S90513	Symphysodon	aequifasciatus	DISCUS "FLÄCHIG BLAU/TÜRKIS / SOLID BLUE/TURQUOISE"	129
S90514	Symphysodon	aequifasciatus	DISCUS "FLÄCHIG BLAU/TÜRKIS / SOLID BLUE/TURQUOISE"	129
S90515	Symphysodon	aequifasciatus	DISCUS "FLÄCHIG BLAU/TÜRKIS / SOLID BLUE/TURQUOISE"	130
S90516	Symphysodon	aequifasciatus	DISCUS "FLÄCHIG BLAU/TÜRKIS / SOLID BLUE/TURQUOISE"	130
S90517	Symphysodon	aequifasciatus	DISCUS "FLÄCHIG BLAU/TÜRKIS / SOLID BLUE/TURQUOISE"	130
S90518	Symphysodon	aequifasciatus	DISCUS "FLÄCHIG BLAU/TÜRKIS / SOLID BLUE/TURQUOISE"	130
S90519	Symphysodon	aequifasciatus	DISCUS "FLÄCHIG BLAU/TÜRKIS / SOLID BLUE/TURQUOISE"	130
S90520	Symphysodon	aequifasciatus	DISCUS "FLÄCHIG BLAU/TÜRKIS / SOLID BLUE/TURQUOISE"	130
S90522	Symphysodon	aequifasciatus	DISCUS "FLÄCHIG BLAU/TÜRKIS / SOLID BLUE/TURQUOISE"	131
S90524	Symphysodon	aequifasciatus	DISCUS "FLÄCHIG BLAU/TÜRKIS / SOLID BLUE/TURQUOISE"	131
S90525	Symphysodon	aequifasciatus	DISCUS "FLÄCHIG BLAU/TÜRKIS / SOLID BLUE/TURQUOISE" "Reflection-Discus"	131
S90527	Symphysodon	aequifasciatus	DISCUS "FLÄCHIG BLAU/TÜRKIS / SOLID BLUE/TURQUOISE"	10/132
S90528	Symphysodon	aequifasciatus	DISCUS "FLÄCHIG BLAU/TÜRKIS / SOLID BLUE/TURQUOISE"	10/132
S90529	Symphysodon	aequifasciatus	DISCUS "FLÄCHIG BLAU/TÜRKIS / SOLID BLUE/TURQUOISE"	10/132
S90530	Symphysodon	aequifasciatus	DISCUS "FLÄCHIG BLAU/TÜRKIS / SOLID BLUE/TURQUOISE"	10/132
S90531	Symphysodon	aequifasciatus	DISCUS "FLÄCHIG BLAU/TÜRKIS / SOLID BLUE/TURQUOISE"	10/132
S90532	Symphysodon	aequifasciatus	DISCUS -SOLID TURQUOISE	128
S90533	Symphysodon	aequifasciatus	DISCUS "FLÄCHIG BLAU/TÜRKIS / SOLID BLUE/TURQUOISE"	133
S90534	Symphysodon	aequifasciatus	DISCUS "FLÄCHIG BLAU/TÜRKIS / SOLID BLUE/TURQUOISE"	133
S90535	Symphysodon	aequifasciatus	DISCUS "FLÄCHIG BLAU/TÜRKIS / SOLID BLUE/TURQUOISE"	133
S90536	Symphysodon	aequifasciatus	DISCUS "FLÄCHIG BLAU/TÜRKIS / SOLID BLUE/TURQUOISE"	133
S90537	Symphysodon	aequifasciatus	DISCUS "FLÄCHIG BLAU/TÜRKIS / SOLID BLUE/TURQUOISE"	133
S90538	Symphysodon	aequifasciatus	DISCUS "FLÄCHIG BLAU/TÜRKIS / SOLID BLUE/TURQUOISE"	133/35
S90539	Symphysodon	aequifasciatus	DISCUS "FLÄCHIG BLAU/TÜRKIS / SOLID BLUE/TURQUOISE" "Blue Diamond"	134
S90540	Symphysodon	aequifasciatus	DISCUS "FLÄCHIG BLAU/TÜRKIS / SOLID BLUE/TURQUOISE" "Blue Diamond"	134
S90541	Symphysodon	aequifasciatus	DISCUS "FLÄCHIG BLAU/TÜRKIS / SOLID BLUE/TURQUOISE" "Blue Diamond"	134
S90542	Symphysodon	aequifasciatus	DISCUS "FLÄCHIG BLAU/TÜRKIS / SOLID BLUE/TURQUOISE" "Blue Diamond"	134
S90543	Symphysodon	aequifasciatus	DISCUS "FLÄCHIG BLAU/TÜRKIS / SOLID BLUE/TURQUOISE" "Blue Diamond"	134
S90544	Symphysodon	aequifasciatus	DISCUS "FLÄCHIG BLAU/TÜRKIS / SOLID BLUE/TURQUOISE" "Blue Diamond"	134
S90545	Symphysodon	aequifasciatus	DISCUS "FLÄCHIG BLAU/TÜRKIS / SOLID BLUE/TURQUOISE" "Blue Diamond"	136
S90546	Symphysodon	aequifasciatus	DISCUS "FLÄCHIG BLAU/TÜRKIS / SOLID BLUE/TURQUOISE" "Blue Diamond"	136
S90547	Symphysodon	aequifasciatus	DISCUS -SOLID TURQUOISE	128
S90548	Symphysodon	aequifasciatus	DISCUS "FLÄCHIG BLAU/TÜRKIS / SOLID BLUE/TURQUOISE" "Blue Diamond"	136
S90549	Symphysodon	aequifasciatus	DISCUS -SOLID TURQUOISE	128
S91501	Symphysodon	aequifasciatus	DISCUS GRÜN/GREEN "COARI" F1	97
S91502	Symphysodon	aequifasciatus	DISCUS GRÜN/GREEN "COARI" F1	98
S91503	Symphysodon	aequifasciatus	DISCUS GRÜN/GREEN "COARI" F1	98
S91504	Symphysodon	aequifasciatus	DISCUS GRÜN/GREEN "COARI" F1	97
S91505	Symphysodon	aequifasciatus	DISCUS GRÜN/GREEN "RED SPOTTED"	99
S91506	Symphysodon	aequifasciatus	DISCUS GRÜN/GREEN "RED SPOTTED"	99
S91507	Symphysodon	aequifasciatus	DISCUS GRÜN/GREEN "RED SPOTTED"	99
S91508	Symphysodon	aequifasciatus	DISCUS GRÜN/GREEN "RED SPOTTED"	99
S91509	Symphysodon	aequifasciatus	DISCUS GRÜN/GREEN "RED SPOTTED"	99
S91510	Symphysodon	aequifasciatus	DISCUS GRÜN/GREEN "RED SPOTTED"	99
S91511	Symphysodon	aequifasciatus	DISCUS GRÜN/GREEN "RED SPOTTED"	100
S91512	Symphysodon	aequifasciatus	DISCUS GRÜN/GREEN "RED SPOTTED"	100
S91513	Symphysodon	aequifasciatus	DISCUS GRÜN/GREEN "RED SPOTTED"	100
S91514	Symphysodon	aequifasciatus	DISCUS GRÜN/GREEN "RED SPOTTED"	100
S91515	Symphysodon	aequifasciatus	DISCUS GRÜN/GREEN "RED SPOTTED"	100
S91516	Symphysodon	aequifasciatus	DISCUS GRÜN/GREEN "RED SPOTTED"	100/169
S91520	Symphysodon	aequifasciatus	DISCUS GRÜN/GREEN "RED SPOTTED" F5	101
S91521	Symphysodon	aequifasciatus	DISCUS GRÜN/GREEN "RED SPOTTED" F5	101
S91522	Symphysodon	aequifasciatus	DISCUS GRÜN/GREEN "RED SPOTTED" F5	101
S91523	Symphysodon	aequifasciatus	DISCUS GRÜN/GREEN "RED SPOTTED" F5	101
S91524	Symphysodon	aequifasciatus	DISCUS GRÜN/GREEN "RED SPOTTED" F5	101
S91525	Symphysodon	aequifasciatus	DISCUS GRÜN/GREEN "RED SPOTTED" F5	101
S91526	Symphysodon	aequifasciatus	DISCUS GRÜN/GREEN "RED SPOTTED" F5	101
S91527	Symphysodon	aequifasciatus	DISCUS GRÜN/GREEN "RED SPOTTED" F5	101
S91528	Symphysodon	aequifasciatus	DISCUS GRÜN/GREEN "RED SPOTTED" F5	101
S91529	Symphysodon	aequifasciatus	DISCUS GRÜN/GREEN "RED SPOTTED" F5	101
S91530	Symphysodon	aequifasciatus	DISCUS GRÜN/GREEN "RED SPOTTED" F5	101
S91532	Symphysodon	aequifasciatus	DISCUS GRÜN/GREEN "RED SPOTTED" F5	103
S91533	Symphysodon	aequifasciatus	DISCUS GRÜN/GREEN "RED SPOTTED" F5	103
S91535	Symphysodon	aequifasciatus	DISCUS GRÜN/GREEN "RED SPOTTED" F5	104

Cichliden- Verbände Weltweit
Cichlid Associations Worldwide

Australia
Queensland Cichlid Group
P.O. Box 380
Wooloongabba, Queensland 4102

Victorian Cichlid Society
23 Mangana Drive
Mulgrave, Victoria 3170

Austria
Deutsche Cichliden Gesellschaft
Victor Kaplan Str. 1-9/1/3/12
A-1220 Wien

Belgium
Belgische Cichliden Vereninging
Kievitlaan 23
B-2288 Ransl

Denmark
Dansk Cichlide Selskab
Tølløsevej 76.
DK-2700 Brønshøj

France
Association France Cichlid
15 Rue des Hirondelles
F-87350 Dauendorf

Germany
Deutsche Cichliden Gesellschaft
Parkstr. 21a
D-33719 Bielefeld

Aquaristischer Arbeitskreis Leinetal
Interessengemeinschaft Cichliden
Ludwig-Prandl-Str. 66
D-37077 Göttingen

Cichliden-Freunde Viernheim
Am Pfarrgarten 12
D-68519 Viernheim

Cichlidenklub Essen
Lohstr. 39
D-45359 Essen

VDA-Arbeitskreis Zwergcichliden
Richard-Holz-Str. 4
D-08060 Zwickau

Hungary
Sügérbarátok Klubja Budapest
Mészöly u. 6. II/3
H-1117 Budapest

Italy
Associazone Italiana Ciclidofili
Via Zucchini, 6
I-48018 Faenza

Japan
Japan Cichlid Association
Kuboyama 1-36-5-A201
Tokyo 192

Mexico
Grupo Mexicano de Cicliófilos
Cordillera Karakorum 223B
Lomas 3a secclón
San Luis Potosl. S.L.P.. 78216

Netherlands
Nederlandse Cichliden Vereniging
Boeier 31
NL-1625 CJ Hoorn

Slovakia
SZCH Klub Chovatelov Cichlid
Prikopova 2
83103 Bratislava

Sweden
Nordiska Ciklid Säliskapet
Plommonvägen 26
S-161 52 Bromma

Switzerland
Deutsche Cichliden Gesellschaft
Am Balsberg 1
CH-8302 Kloten

Taiwan (R.O.C.)
Taiwanese Cichlid Association
N°17, Lana 239, An-Ho Road
Taipei

United Kingdom
Britsh Cichlid Association
248 Longridge, Knutsford
Cheshire, WA 18 8PH

U.S.A.
American Cichlid Association
P.O. Box 5351
Naperville, IL 60587-5351

Adv. Cichl. Aquarists South California
P.O. Box 8173
San Marino, CA 91108

African Cichlid Club
3744 Forest Valley Court SE
Grand Rapids, MI 49508

Apistogramma Study Group
3724 N. Mobile
Chicago, IL 60646

Beach Citles Cichlid Assiciation
2106 Manhattan Beach Boulevard/5
Redondo Beach, CA 90278

Cichlasoma Study Group
1813 Locks Mill Dr.
Fenton, Mo. 63026

Cichlasoma Study Group
1813 Locks Mill Dr.
Fenton, MO 63026

Cichlid Hobbyists Eastern Wisconsin
3259 So. Swain Court
Milwaukee, WI 53204

Cichlid Seekers
2014 45th Street Court N.W.
Big Harbor, Washington 98335

Fort Wayne Cichlid Association
9638 Manor Woods Rdl.
Ft. Wayne, IN 46804

Greater Chicago Cichlid Association
2633 N. Rhodes
River Grove, IL 60171

Greater Cincinnati Cichlid Association
15 W. Southern Avenue
Covington, KY 41015

Illinois Cichlids and Scavengers
7807 Sunset Drive
Elmwood Park, IL 60635

Lake Erie Cichlid Association
1113 Sunset Road
Mayfield Heights, OH 44124

Michigan Cichlid Association
P.O. Box 59
New Baltimore, MI 48047

Milwaukee Cichlid Club
1926 Grange Avenue
Racine, WI 53403

Ohio Cichlid Association
7330 Arnes Road
Parma, OH 44129

Oregon Cichlid Study Group
388 N. State Street
Lake Oswego, OR 97034

Pacific Coast Cichlid Assiciation
P.O. Box 28145
San Jose, CA 95128

Pikes Peak Cichlid Association
P.O. Box 17176
Colorado Springs, CO 80935

Rift Valley Cichlids
15800 Laguna Avenue
Lake Elsinore, CA 92530

Rocky Mountain Cichlid Association
5065 W. Hinsdale Cir.
Littleton, CO 80123

South American Cichlid Study Group
P.O. Box 14123
Dinkytown Station,
Minneapolis, MN 55414-0123

Southern California Cichlid Association
P.O. Box 574
Midway City, CA 92655

Irrtum vorbehalten.
Sämtliche Angaben ohne Gewähr.
Die Liste erhebt keinen Anspruch auf
Vollständigkeit.

Bilder:

Wir danken den nachfolgend aufgelisteten Spezialisten und Firmen für die freundliche Überlassung ihrer Dias und für ihre Beratung, auch denen, die wir eventuell vergessen haben zu erwähnen.

Photographs:

We would like to express our gratitude to the following specialists and companies for kindly offering us their slides and for their advice. Our thanks also go to those whose names we may have forgotten.

Godwin K.M. Sim	**M. Göbel**
M. Tomzana	**F. Bodenmüller**
H.-J. Mayland	**J. Schütz**
H. Morche	**W. Mikschofsky**
M. Smith	**Fa. Zoo-Zajac**
U. Werner	**H. Colle**
A. Canovas	**W. Konrad**
F. Teigler	**J. Stendker**
B. Migge	**W. A. Tomey**
H. Reinhard	**Wayne D.C. HongKong**
F. Schmidt-Knaatz	**TFF Mainland**
H. Linke	**Takrit Aquarium, Bangkok**
J. Glaser	**Ch. W. Shing**
H. Nakano	**F. Schulten**
F. Schäfer	**Dr. L. Y. Hoe**
I. den Daas	**L. T. Juan, Tl.**
J. Dawes	**Sea View Aquarium**
E. Schraml	
P. Schlingmann	

Aquarium Glaser GmbH,
die uns von ihren wöchentlichen Importen immer fotogene Tiere zur Verfügung stellten.

for providing photogenic fish from their weekly imports.

amtra - Aquaristic GmbH,
für die zur Verfügung gestellten Fotobecken
for the aquaria for photography.

Tierärztliche Beratung:
Veterinary consulting
Dr. Markus Biffar,
Fach-Tierarzt + Spezialist für Fischmedizin
veterinary surgeon (fish specialist)

Ausführlichere Informationen über Pflege und Zucht der Fische finden Sie in Fachbüchern, Zeitschriften und der ersten und einzigen internationalen Zeitung für Aquarianer, der AQUALOGnews. In dieser Zeitung werden darüber hinaus neue und neuentdeckte Arten, Varianten und Zuchtformen als Einklebebilder - sogenannte Stickups - veröffentlicht. Diese Einklebebilder halten Ihren AQUALOG auf Jahre hinaus aktuell.

Detailed information on fish care and breeding can be found in specialist books, periodicals and in the one and only international newspaper for aquarists, the AQUALOGnews. In this newspaper we also publish newly discovered or imported species, varieties or breeding forms as stickers: the so-called Stickups. With these stickers you can keep your AQUALOG up-to-date for years and years.

Cover Photos

Frontcover:

Grüner Diskus (photo: H.-J. Mayland)

Koi-Skalar, Schleierskalar (photos: F. Teigler/ ACS)

Backcover:

Goldkopf-Skalar (photo: Nakano/ACS)

Rot-türkis Diskus (photo: Wayne DC, HongKong)

Diskus rot-geperlt (photo: Wayne DC, HongKong)

Diskus der offenen Klasse (photo: H.-J. Mayland)

Symbols

In order to include as many pictures as possible, and bearing the international nature of the publication in mind, we have intentionally decided against detailed textual descriptions, replacing them by international symbols. This way, one can easily obtain the most important facts about the species and its care.

Continent of origin:

simply check the letter in front of the code-number
A = Africa E = Europe + North America
S = South America X = Asia + Australia

Age:

the last number of the code always stands for the age of the fish in the photo:

1 = small   (baby, juvenile colouration)
2 = medium   (young fish / saleable size)
3 = large   (half-grown / good saleable size)
4 = XL   (fully grown / adult)
5 = XXL   (breeder)
6 = show   (show-fish)

Immediate origin:

W = wild
B = bred
Z = breeding-form
X = cross-breed

Size:

..cm = approximate size these fish can reach as adults.

Sex:

♂ male   ♀ female   ♂♀ pair

Temperature:

◁ 18-22°C (68 - 72°F) (room-temperature)
▷ 22-25°C (71 - 77°F) (tropical fish)
△ 24-29°C (75 - 85°F) (Discus etc)
▽ 10-22°C (50 - 72°F) cold

pH-Value:

₽   pH 6,5 - 7,2 no special requirements (neutral)
↓P pH 5,8 - 6,5 prefers soft, slightly acidic water
↑P ph 7,5 - 8,5 prefers hard, alkaline water

Lighting:

○ bright, plenty of light / sun
◐ not too bright
● almost dark

Food:

☺ omnivorous / dry food, no special requirements
☻ food specialist, live food/ frozen food
☹ predator, feed with live fish
☯ plant-eater, supplement with plant food

Swimming:

⊞ no special characteristics
⊟ in upper area / surface fish
⊟ in lower area / floor fish

Aquarium- set up:

▭ only floor and stones etc.
▧ stones / roots / crevices
▨ plant aquarium + stones / roots

Behaviour / reproduction:

♥ keep a pair or a trio
🐟 school fish, do not keep less than 10
🐠 egg-layer
🐡 livebearers / viviparous
🐟 mouthbrooder
⬭ cavebrooder
🦐 bubblenest-builder
◯ algae-eater / glass-cleaner (roots + spinach)
◇ non aggressive fish, easy to keep (mixed aquarium)
⚠ difficult to keep, read specialist literature beforehand
🛑 warning, extremely difficult, for experienced specialists only
❶ the eggs need a special care
§ protected species (WA), special license required ("CITES")

Minimum tank: capacity:

ss	super small	20 - 40 cm	5 - 20 l
s	small	40 - 80 cm	40 - 80 l
m	medium	60 - 100 cm	80 - 200 l
l	large	100 - 200 cm	200 - 400 l
xl	XL	200 - 400 cm	400 - 3000 l
xxl	XXL	over 400 cm	over 3000 l
			(show aquarium)

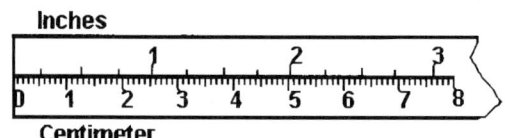

240   *South American* **Cichlids IV**   © **V e r l a g   A . C . S .   G m b H**